Current Topics in Microbiology
108 and Immunology

Editors

M. Cooper, Birmingham/Alabama · W. Goebel, Würzburg
P.H. Hofschneider, Martinsried · H. Koprowski, Philadelphia
F. Melchers, Basel · M. Oldstone, La Jolla/California
R. Rott, Gießen · H.G. Schweiger, Ladenburg/Heidelberg
P.K. Vogt, Los Angeles · R. Zinkernagel, Zürich

Methylation of DNA

Edited by Thomas A. Trautner

With 22 Figures

Springer-Verlag
Berlin Heidelberg New York Tokyo 1984

Professor Dr. Thomas A. Trautner
Max-Planck-Institut
für molekulare Genetik
Ihnestraße 63/73
D-1000 Berlin 33

ISBN 3-540-12849-2 Springer-Verlag Berlin Heidelberg New York Tokyo
ISBN 0-387-12849-2 Springer-Verlag New York Heidelberg Berlin Tokyo

© by Springer-Verlag Berlin Heidelberg 1984.
Library of Congress Catalog Card Number 15-12910
Printed in Germany.

The use of registered names, trademarks, etc. in this publication does not imply,
even in the absence of a specific statement, that such names are exempt from
the relevant protective laws and regulations and therefore free for general use.

Product Liability: The publisher can give no guarantee for information about
drug dosage and application thereof contained in this book. In every individual
case the respective user must check its accuracy by consulting other pharmaceuti-
cal literature.

Typesetting, printing and bookbinding:
Universitätsdruckerei H. Stürtz AG, Würzburg
2123/3130-543210

Preface

Postreplicative methylation of bacterial DNA has long been known to be the molecular basis of "modification," which protects DNA against destruction by restriction endonucleases. More recently, another function of DNA methylation was found in *Escherichia coli,* where methylation is involved during DNA replication in the recognition of old and newly synthesized strands. The intensive search for new restriction enzymes during the 1970s yielded an enormous arsenal of such enzymes and revealed the ubiquitous distribution of restriction/modification systems in the bacterial kingdom without providing much information on the corresponding modification methyltransferases. However, it is obvious that DNA methyltransferases represent an ideal class of enzymes to those interested in protein/DNA interactions; these enzymes are at least as interesting as the restriction enzymes, with which they share the capacity to recognize and interact with specific sequences of DNA.

In recent years the interest in DNA methylation has been greatly stimulated by two discoveries: the correlation between gene expression and hypomethylation in eukaryotes and the convertability of DNA into its Z form through cytosine methylation. In fact, studies on DNA methylation are now being intensively performed in many laboratories. A description of the state of the art of DNA methylation has been the topic of two congresses: The Cologne Spring Meeting in 1981 organized by WALTER DOERFLER and an EMBO Workshop at Nethybridge in 1982 organized by ROGER ADAMS.

Due to the expansion of this field, it is no longer feasible for a review book to be encyclopedic in dealing with DNA methylation. However, this volume does contain a number of topical reviews which will, hopefully, be of broad interest. I wish to thank the contributors of the volume for their cooperation and the members of the staff of Springer-Verlag for their patient and meticulous editing of articles. I am grateful to my colleague

DR. URSULA GÜNTHERT for many suggestions and discussions during the editing of this volume.

Berlin, January 1984   THOMAS A. TRAUTNER

Table of Contents

Indexed in Current Contents

List of Contributors

ADAMS, R.L.P., Department of Biochemistry, University of Glasgow, Glasgow G23 8QQ, United Kingdom

BICKLE, T.A., Microbiology Department, Biozentrum, University of Basel, Klingelbergstraße 70, CH-4056 Basel

BIRD, A.P., MRC Mammalian Genome Unit, King's Buildings, West Mains Road, Edinburgh EH9 3JT, United Kingdom

BURDON, R.H., Department of Biochemistry, University of Glasgow, Glasgow G23 8QQ, United Kingdom

CHRISTMAN, J.K., Departments of Biochemistry and Pediatrics, Mount Sinai School of Medicine, 1 Gustave Levy Plaza, New York, NY 10029, USA

CONSTANTINIDES, P.A., Department of Medical Biochemistry, University of Cape Town, Observatory 7925, South Africa

DAVIS, T., Department of Biochemistry, University of Glasgow, Glasgow G23 8QQ, United Kingdom

DOERFLER, W., Institute of Genetics, University of Cologne, Weyertal 121, D-5000 Köln 41

FULTON, J., Department of Biochemistry, University of Glasgow, Glasgow G23 8QQ, United Kingdom

GRABOWY, T., Dana-Farber Cancer Institute, Division of Cancer Genetics, 44 Binney Street, Boston, MA 02115, USA

GÜNTHERT, U., Max-Planck-Institut für Molekulare Genetik, Ihnestraße 63–73, D-1000 Berlin 33 (Dahlem)

JONES, P.A., Departments of Pediatrics and Biochemistry, University of Southern California, 4650 Sunset Boulevard, Los Angeles, CA 90027, USA

KAHMANN, R., Max-Planck-Institut für Molekulare Genetik, Ihnestraße 63–73, D-1000 Berlin 33

KIRK, D., Department of Biochemistry, University of Glasgow, Glasgow G23 8QQ, United Kingdom

NAGARAJA, V., Microbiology Department, Biozentrum, University of Basel, Klingelbergstraße 70, CH-4056 Basel

QURESHI, M., Department of Biochemistry, University of Glasgow, Glasgow G23 8QQ, United Kingdom

RADMAN, R., Institut Jacques Monod, C.N.R.S., Université Paris VII, Tour 43, 2 Place Jussieu, F-75251 Paris Cedex

SAGER, R., Dana-Farber Cancer Institute, Division of Cancer Genetics, 44 Binney Street, Boston, MA 02115, USA

SANO, H., Dana-Farber Cancer Institute, Division of Cancer Genetics, 44 Binney Street, Boston, MA 02115, USA

SURI, B., Microbiology Department, Biozentrum, University of Basel, Klingelbergstraße 70, CH-4056 Basel

TAYLOR, S.M., Division of Hematology-Oncology, Childrens Hospital of Los Angeles, University of Southern California, 4650 Sunset Boulevard, Los Angeles, CA 90027, USA

TRAUTNER, T.A., Max-Planck-Institut für Molekulare Genetik, Ihnestraße 63–73, D-1000 Berlin 33 (Dahlem)

VANYUSHIN, B.F., A.N. Belozersky Laboratory of Molecular Biology and Bioorganic Chemistry, Moscow State University, Moscow 117234, USSR

WAGNER, R., Institut Jacques Monod, C.N.R.S., Université Paris VII, Tour 43, 2 Place Jussieu, F-75251 Paris Cedex

Bacterial DNA Modification

B. Suri, V. Nagaraja, and T.A. Bickle

1 Introduction

As first proposed by Arber (1965), DNA restriction/modification systems (R/M systems) are mediated by endonucleases and DNA methylases that recognize the same DNA sequences. The endonuclease recognizes its specific sequence as a signal to cleave the DNA unless the sequence has been previously methylated by the modification enzyme. Chromosomal DNA from cells harboring the R/M system is normally methylated, and is thus not a substrate for the restriction enzyme. Foreign DNA lacking the specific methylation pattern and introduced into the cell by phage infection, conjugation, or transformation is the only known natural substrate for restriction. R/M systems can therefore be considered primitive prokaryotic analogues of the eukaryotic immune system.

A vast body of literature on the genetics and biochemistry of R/M systems has accumulated since they were first investigated 20 years ago (Arber and Dussoix 1962), and it is now clear that R/M systems can be conveniently classified into three types (Boyer 1971; Kauc and Piekaro-wicz 1978; Nathans and Smith 1975).

The most complicated of these are the type-I systems, which are mediated by complex, multifunctional enzymes and which were the first proteins shown to recognize specific DNA sequences. The restriction enzymes EcoK and EcoB from the *Escherichia coli* strains K12 and B are the two prototypes, and are still the only ones to have been studied in detail.

Microbiology Department Biozentrum, University of Basel, Klingelbergstrasse 70, CH-4056 Basel

Current Topics in Microbiology and Immunology, Vol. 108
© Springer-Verlag Berlin · Heidelberg 1984

The enzymes contain three nonidentical subunits coded by three contiguous genes on the *E. coli* chromosome. They require Mg^{+2}, ATP, and *S*-adenosylmethionine (AdoMet) for enzymatic activity and can function as restriction endonucleases (cleaving the DNA far from their recognition sequence), ATPases, or, as detailed later, modification methylases.

The type-II R/M systems include all those enzymes that have simple cofactor requirements (Mg^{+2} for restriction, AdoMet for modification) and simple subunit structures, and in which the restriction enzymes cut the DNA at, or close to, the sequences that they recognize. In the cases that have been investigated, separate enzymes catalyze restriction and modification. The type-II restriction enzymes are those that have found such wide application in recent years in molecular genetics.

Type-III R/M systems have been characterized more recently than the others, and represent an intermediate level of complexity. The restriction enzymes contain two nonidentical subunits and require ATP and Mg^{+2} for activity. AdoMet is not required for the endonuclease activity, as it is by type-I enzymes, although it stimulates the reaction. In the presence of both AdoMet and ATP, restriction and modification methylation are competing reactions. A separate modification enzyme which consists of the smaller of the two subunits of the restriction enzyme has also been isolated. Detailed reviews on restriction enzymes have recently been published (BICKLE 1982; MODRICH 1979; NATHANS and SMITH 1975; YUAN 1981; MODRICH and ROBERTS 1982). Here we will restrict ourselves to modification methylases, with emphasis on type-I systems.

2 Various Modification Methylases

Modification methylases methylate specific residues within their recognition sequences at either the ^{6}N position of adenine or the ^{5}C position of cytosine, depending on the system. They use AdoMet as the methyl donor and, once methylated, the DNA is resistant to cleavage by the corresponding restriction enzyme.

Although type-I restriction enzymes can modify appropriate substrate DNAs, separate modification methylases can be isolated. The modification enzymes from *E. coli* B (LAUTENBERGER and LINN 1972a) and *E. coli* K12 (this paper) have been characterized. They both contain two nonidentical subunits which are the same as two of the three subunits of the corresponding restriction enzymes.

The few type-II modification enzymes that have been characterized have proved to be relatively simple enzymes. One of them is discussed in detail elsewhere in this volume (U. GÜNTHERT and T.A. TRAUTNER). In general, they require only a substrate DNA and AdoMet for activity (DU-GAICZYK et al. 1974; RUBIN and MODRICH 1977). Although both the restriction enzymes and the modification methylases recognize the same DNA sequences, they seem to be physically and genetically unrelated.

A modification enzyme lacking endonuclease activity was isolated from the type-III *Eco*P1 system as early as 1972 (BROCKES et al. 1972). Detailed studies on the type-III modification methylases, however, have only been done recently in this laboratory (HADI et al. 1983; IIDA et al. 1983). The modification methylase consists of one of the two subunits of the restriction enzyme, this subunit being responsible for recognition of the specific DNA sequences in both the restriction and the modification reactions. Unlike most modification methylases, this one requires Mg^{+2} as well as AdoMet for methylation. In contrast to modification by the restriction enzyme, ATP does not stimulate the reaction.

3 Features of Modification Sequences

A general feature of most recognition sequences for restriction and modi-fication enzymes is that they have methylatable residues in both strands of the DNA. This is extremely important physiologically because DNA in which only one strand carries the specific methylation (hemimethylated DNA) is resistant to cleavage by the corresponding restriction enzyme. Since hemimethylated DNA is the normal product of DNA replication or repair, this feature of the reaction provides the mechanism whereby cells avoid restricting their own chromosomal DNA. For the type-I enzymes *Eco*K and *Eco*B, hemimethylated DNA is by far the preferred substrate for methylation, the reaction rate with this substrate being more than 100 times faster than with completely nonmodified DNA (*Vovis* et al. 1974; BURCKHARDT et al. 1981). In contrast, the type-II *Eco*RI methylase shows no preference for hemimethylated over nonmethylated sites (RUBIN and MODRICH 1977).

Some interesting exceptions can be found to the rule that fully modi-fied R/M recognition sites are methylated in both strands of the DNA. All three of the type-III enzymes known at present methylate adenosyl residues in one strand of the DNA only. Two of the recognition se-quences have no adenosyl residues in the other strand, and while the other – that of *Hin*fIII – has adenosyl residues in both strands, only one strand is methylated (BÄCHI et al. 1979; HADI et al. 1979; PIEKAROWICZ et al. 1981). In cells carrying these R/M systems, DNA replication gener-ates one daughter DNA molecule containing the parental modification and a second daughter with the corresponding recognition site completely unmodified. These unmodified sites ought to be targets for restriction and we do not yet understand how cells containing type-III R/M systems avoid restricting their own chromosomal DNA.

Some of the type-II R/M recognition sequences are asymmetric and are most likely only methylatable in one strand. *Mbo*II is one such exam-ple. The recognition sequence of this enzyme is 5'-GAAGA-3' in one strand and 5'-TCTTC-3' in the other: one strand contains no cytosines and the other no adenosines (BROWN et al. 1980). Unless the *Mbo*II modification methylase is capable of methylating both adenosyl and cytosyl residues, a property that has not yet been found for any DNA methylase, only one strand can be methylated in modified DNA (Bächi et al. 1979).

4 Type-I Modification Methylases

4.1 Genetic Organization

The type-I R/M systems of *E. coli* K12 and B are fully specified by three contiguous genes mapping at 98.5 min on the *E. coli* K12 chromosome (Sain and Murray 1980; Bachmann and Low 1980). Genetic, physical, and immunochemical studies have revealed that these two systems are allelic and that they also share homology with several R/M systems from different *Salmonella* species (Boyer and Roulland-Dussoix 1969; Bullas et al. 1980; Hubacek and Glover 1970; Murray et al. 1982). Recent studies have shown that another system which, in physiological studies, behaves as though it were type I (Lark and Arber 1970) – the *Eco*A system of *E. coli* 15T⁻ (Arber and Wauters-Willems 1970) – in fact shows no homology on the DNA level with the classical type-I systems. Moreover, antibodies prepared against *Eco*K do not cross-react with extracts from cells expressing *Eco*A (Murray et al. 1982).

The three genes involved in the *E. coli* K12 and B R/M systems are called *hsd*R, *hsd*M, and *hsd*S (*hsd*, for "host specificity for DNA"). Strains carrying mutations in the *hsd*R gene are defective in restriction, those with mutated *hsd*S genes lack both restriction and modification, and *hsd*M mutants can be isolated only in strains that already carry a mutation in one of the other two genes. These phenotypes, together with the results of complementation analysis, led to the suggestion that all three gene products are necessary for restriction, while the *hsd*M and *hsd*S gene products suffice for modification (Boyer and Roulland-Dussoix 1969; Hubacek and Glover 1970). The *hsd*S gene product would be responsible for recognizing the specific sequences in DNA in both restriction and modification, the *hsd*M gene product would catalyze modification, and both *hsd*M and *hsd*R would be required for restriction.

The *hsd* locus of *E. coli* K12 has been cloned in bacteriophage λ and a deletion analysis of this cloned DNA has revealed the gene order to be *hsd*R-*hsd*M-*hsd*S. All three genes are transcribed in the same direction but from two promoters, one of them upstream of *hsd*R and the other between *hsd*R and *hsd*M (Sain and Murray 1980). This organization of the *hsd* locus into two transcriptional units allows the genes which code for the subunits of the modification enzyme to be transcribed independently of the *hsd*R gene, and this may be of considerable physiological importance.

4.2 Enzyme Structure

Before the discovery that the type-I restriction enzymes, which contain the products of all three *hsd* genes, had modification activity (VOVIS et al. 1974), it was predicted that the type-I modification methylases ought to contain the products of the *hsd*S and *hsd*M genes only. Such an enzyme was isolated 10 years ago from *E. coli* B (LAUTENBERGER and LINN 1972a). The enzyme contained two nonidentical subunits with molecular weights of 60000 and 55000 which comigrated on polyacrylamide-SDS gels with the two smaller subunits of the *Eco*B restriction enzyme (LAU-TENBERGER and LINN 1972b). The stoichiometry of the subunits in the enzyme was variable and changed upon storage. However, freshly purified enzyme had equimolar amounts of the two subunits.

Quite recently we isolated a modification methylase from an *E. coli* K12 strain in which the *hsd* genes are transcribed from the strong bacteriophage λ promoter, P_L (MURRAY et al. 1982). This enzyme also contains the *hsd*M and S gene products in equimolar amounts. Its reaction characteristics will be described in the next section.

4.3 Kinetics of Methylation

We have compared the methylation properties of the *Eco*K restriction enzyme with those of the modification methylase described in the preceding section. With unmodified DNA, the restriction enzyme is considerably less effective than the methylase and is partially inhibited by the presence of ATP. The methylase is unaffected by the presence of ATP (Fig. 1). For both enzymes the overall reaction is of the first order, indicating that the rate-limiting step follows the formation of the enzyme-DNA complex. The first-order rate constants were $6 \times 10^{-5}\,\mathrm{s}^{-1}$ for the methylase and $2 \times 10^{-5}\,\mathrm{s}^{-1}$ for the restriction enzyme (without ATP). The apparent inhibition by ATP of the reaction with the restriction enzyme may be due to the fact that ATP stimulates a conformational change in the enzyme, which was shown to result in the release of AdoMet bound to it (BICKLE et al. 1978). Results similar to these have been obtained for *Eco*B and the B-specific methylase (HABERMAN et al. 1972; LAUTENBERGER and LINN 1972a).

The kinetics of modification of hemimethylated pBR322 DNA (prepared by hybridizing modified with nonmodified DNA) is shown in Fig. 2. The reaction is now much faster with both enzymes; (the time scale of Fig. 2 is minutes, whereas that of Fig. 1 is hours). For the restriction enzyme, the results are similar to those previously reported for *Eco*B and *Eco*K (VOVIS et al. 1974; BURCKHARDT et al. 1981). The principal difference is that the stimulation of the reaction by ATP and Mg^{+2} reported earlier is now seen to be composed of about a 1.6-fold stimulation by Mg^{+2} and a further twofold stimulation by ATP. The modification enzyme is slightly more efficient than the restriction enzyme and is unaffected by the presence of ATP. For both enzymes, the reaction is again

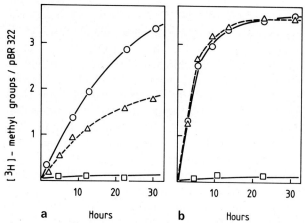

Fig. 1a, b. Methylation of unmodified DNA, **a** by the *Eco*K restriction enzyme, **b** by the modification enzyme. The reactions contained 100 mM 4-(2-hydroxyethyl)-1-piperazine-ethane-sulphonic acid, pH 6.7; 0.25 mM EDTA; 14 mM 2-mercaptoethanol; and 3.6 μM ^{3}H-AdoMet (Amersham, 74 Ci/mmol). When present, Mg^{+2} was at a concentration of 6.6 mM and ATP at 1 mM. The DNA was pBR322, linearized by cleavage with *Sal*I, and was used at a concentration of 7.7 μg/ml. The plasmid pBR322 has two *Eco*K recognition sites. Incubations were at 37° C. Samples were removed at the indicated times, the reaction stopped by the addition of phenol, and the DNA separated from low-molecular-weight radioactive material by gel filtration through small Bio-Gel A 0.5 M columns. The DNA-containing fractions were counted in Instagel (Packard) with an efficiency of 50%. □——□, modified DNA, +ATP; ▵---▵, nonmodified DNA, +ATP; o——o, nonmodified DNA, − ATP

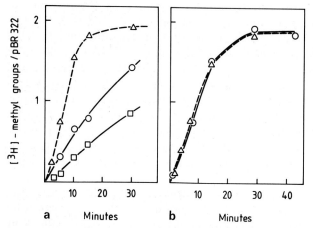

Fig. 2a, b. Methylation of heteroduplex DNA, **a** by the *Eco*K restriction enzyme; **b** by the modification enzyme. The reaction conditions were the same as those described in Fig. 1, except that the reaction mixtures were preheated to 37° for 10 min before adding enzyme, and the DNA concentration was 15 μg/ml. The DNA was prepared by heating equal amounts of modified and nonmodified *Sal*I-linearized pBR322 DNA in a heat-sealed glass capillary at 100° C to melt the DNA; then reannealing was done by allowing the temperature to drop to 55° C over 90 min. The DNA should contain 50% heteroduplex molecules and 25% each of homoduplex modified and nonmodified molecules. The latter are essentially not methylated in the 40 min that this experiment takes (see Fig. 1). ▵---▵, + ATP, + MgCl$_2$; o——o, − ATP, + MgCl$_2$; □——□, − ATP, − MgCl$_2$

of the first order and the first-order rate constants are $2 \times 10^{-3}\,s^{-1}$ for the methylase and $3 \times 10^{-3}\,s^{-1}$ for the restriction enzyme in the presence of both Mg^{+2} and ATP.

A comparison of the rate constants for the modification of nonmodified versus hemimethylated DNA shows that the restriction enzyme methylates the latter some 150 times faster than the former while the corresponding figure for the methylase is 35 times. Under all incubation conditions the methylase is more efficient than the restriction enzyme, especially with nonmodified DNA.

5 Physiological Implications

The most obvious question that arises from these studies on type-I R/M systems is: Why should there be a separate modification methylase when the restriction enzyme present in the same cells is itself an efficient modification methylase? One trivial possibility would be that the methylases are artifacts of the purification procedure. If this were so, it is difficult to see why the hsdS and hsdM genes, which are the structural genes for the subunits of the methylase, should be organized as a single transcriptional unit, while the hsdR gene is transcribed from a separate promoter (SAIN and MURRAY 1980). This arrangement could allow an independent regulation of the production of the restriction and modification enzymes; whether such a regulation occurs is not known.

It might be argued that most bacterial cells are never challenged by foreign DNA, and thus never use the endonuclease activity of their restriction enzymes. A relatively low level of restriction activity may therefore suffice to provide protection to the cells. On the other hand, modification activity is continually required, because DNA replication and repair are constantly generating hemimethylated DNA that must be modified before the next round of replication creates unmodified sites that would be a target for restriction. The relative amounts of restriction and modification enzyme have never been quantitated in *E. coli* strains K12 or B. The availability of antibodies means that such a quantitation is now feasible, and it would be interesting to see whether there is an excess of the modification enzyme.

Very recently, we have purified the enzymes involved in the *Eco*A restriction-modification system. As expected from the earlier studies (MURRAY et al. 1982) described above, these enzymes show some interesting differences from the classic type-I enzymes. For the present argument the most important difference is that the basic enzyme is a modification methylase containing two subunits of approximately the same molecular weights as the classic type-I hsdS and hsdM gene products. A protein of about the same molecular weight as a classic hsdR subunit can be purified separately. This protein has no detectable enzymatic activity by itself; however, when added to the modification methylase in the presence of a substrate DNA the methylase is converted to a restriction endonuclease

(SURI and BICKLE, unpublished results). Thus, cells carrying the *Eco*A restriction-modification system always contain a modification methylase and
probably assemble an active restriction endonuclease only when the cell
is challenged by foreign DNA.

The concentrations of the different cofactors, in particular that of
ATP, may play a crucial role in regulating restriction and modification
in vivo. The restriction enzyme shows an absolute requirement for ATP
in the restriction reaction and is stimulated by it in the modification reaction (with hemimethylated substrates). The modification enzyme, on the
other hand, is unaffected by the presence of ATP. This may ensure that
newly replicated or repaired DNA is efficiently methylated even when
ATP levels in the cell are low. It is worth noting that one condition in
which ATP concentrations are expected to be low is following restriction,
when the restriction enzyme has transformed itself into a potent ATP
hydrolase.

Acknowledgment. Work from this laboratory was supported by grants from the Swiss National Science Foundation.

References

Arber W (1965) Host-controlled modification of bacteriophage. Annu Rev Microbiol
    19:365–377
Arber W, Dussoix D (1962) Host specificity of DNA produced by *Escherichia coli.* I. Host-
    controlled modification of bacteriophage λ. J Mol Biol 5:18–36
Arber W, Wauters-Willems D (1970) Host specificity of DNA produced by *Escherichia coli.*
    XII. The two restriction and modification systems of strain 15T⁻. MGG 108:203–217
Bächi B, Reiser J, Pirrotta V (1979) Methylation and cleavage sequences of the *Eco*P1 restric-
    tion-modification enzyme. J Mol Biol 128:143–163
Bachmann BJ, Low KB (1980) Linkage map of *Escherichia coli* K12, edition 6. Microbiol
    Rev 44:1–56
Bickle TA (1982) The ATP-dependent restriction endonucleases. In: Linn S, Roberts RJ (eds)
    The nucleases. Cold Spring Harbor, New York, pp 85–108
Bickle TA, Brack C, Yuan R (1978) ATP-induced conformational changes in the restriction
    endonuclease from *Escherichia coli* K12. Proc Natl Acad Sci USA 75:3099–3103
Boyer HW (1971) DNA restriction and modification mechanisms in bacteria. Annu Rev Micro-
    biol 25:153–176
Boyer HW, Roulland-Dussoix D (1969) A complementation analysis of the restriction and
    modification of DNA in *Escherichia coli.* J Mol Biol 41:459–472
Brockes JP, Brown PR, Murray K (1972) The deoxyribonucleic acid modification enzyme
    of bacteriophage P1. Biochem J 127:1–10
Brown N, Hutchison III CA, Smith M (1980) The specific nonsymmetrical sequence recognized
    by restriction enzyme *Mbo*II. J Mol Biol 140:143–148
Bullas LR, Colson C, Neufeld B (1980) Deoxyribonucleic acid restriction and modification
    systems in *Salmonella*: Chromosomally located systems of different serotypes. J Bacteriol
    141:275–292
Burckhardt J, Weisemann J, Yuan R (1981) Characterization of the DNA methylase activity
    of the restriction enzyme from *Escherichia coli* K. J Biol Chem 256:4024–4032
Dugaiczyk A, Hedgepeth J, Boyer HW, Goodman HM (1974) Physical identity of the SV40
    deoxyribonucleic acid sequence recognized by the *Eco*RI restriction endonuclease and modi-
    fication methylase. Biochemistry 13:503–512

Haberman A, Heywood J, Meselson M (1972) DNA modification methylase activity of *Escherichia coli* restriction endonucleases K and P. Proc Natl Acad Sci USA 69:3138–3141

Hadi SM, Bächi B, Shepherd JCW, Yuan R, Ineichen K, Bickle TA (1979) DNA recognition and cleavage by the *Eco*P15 restriction endonuclease. J Mol Biol 134:655–666

Hadi SM, Bächi B, Iida S, Bickle TA (1983) DNA restriction-modification enzymes of phage P1 and plasmid P15B: Subunit functions and structural homologies. J Mol Biol 165:19–34

Hubacek J, Glover SW (1970) Complementation analysis of temperature-sensitive host specificity mutations in *Escherichia coli*. J Mol Biol 50:111–127

Iida S, Meyer J, Bächi B, Stålhammar-Carlemalm M, Schrickel S, Bickle TA, Arber W (1983) The DNA restriction-modification genes of phage P1 and plasmid P15B: Structure and in vitro transcription. J Mol Biol 165:1–18

Kauc L, Piekarowicz A (1978) Purification and properties of a new restriction endonuclease from *Haemophilus influenzae* Rf. Eur J Biochem 92:417–426

Lautenberger JA, Linn S (1972a) The deoxyribonucleic acid modification and restriction enzymes of *Escherichia coli* B. I. Purification, subunit structure and catalytic properties of the modification methylase. J Biol Chem 247:6176–6182

Lautenberger JA, Linn S (1972b) The deoxyribonucleic acid modification and restriction enzymes of *Escherichia coli* B. II. Purification, subunit structure and catalytic properties of the restriction endonuclease. J Biol Chem 247:6183–6191

Modrich P (1979) Structures and mechanisms of DNA restriction and modification enzymes. Quart Rev Biophys 12:315–369

Modrich P, Roberts RJ (1982) Type-II restriction and modification enzymes. In: Linn S, Roberts RJ (eds) The nucleases. Cold Spring Harbor, New York, pp 109–154

Murray NE, Gough JA, Suri B, Bickle TA (1982) Structural homologies among type-I restriction-modification systems. EMBO J 1:535–539

Nathans D, Smith HO (1975) Restriction endonucleases in the analysis and restructuring of DNA molecules. Annu Rev Biochem 44:273–293

Piekarowicz A, Bickle TA, Shepherd JCW, Ineichen K (1981) The DNA sequence recognized by the *Hin*fIII restriction endonuclease. J Mol Biol 146:167–172

Rubin RA, Modrich P (1977) *Eco*RI methylase: Physical and catalytic properties of the homogeneous enzyme. J Biol Chem 252:7265–7272

Sain B, Murray NE (1980) The *hsd* (host specificity) genes of *Escherichia coli* K12. MGG 180:35–46

Vovis GF, Horiuchi K, Zinder ND (1974) Kinetics of methylation by a restriction endonuclease from *Escherichia coli* B. Proc Natl Acad Sci USA 71:3810–3813

Yuan R (1981) Structure and mechanism of multifunctional restriction endonucleases. Annu Rev Biochem 50:280–315

DNA Methyltransferases of *Bacillus subtilis* and Its Bacteriophages *

U. GÜNTHERT and T.A. TRAUTNER[1]

1 Introduction

Postreplicative DNA methylation occurs with a high incidence in bacteria and may affect resident DNA and that of infecting bacteriophages. By far the most widespread role of DNA methylation is to provide the DNA with "modification," i.e., protection against the endonucleolytic attack of cellular restriction enzymes. Important aspects of this role in connection with type-I enzymes are discussed in the review of SURI et al. (this volume). Another well-studied physiological function of DNA methylation observed in *Escherichia coli* is its role in strand recognition for proofreading during DNA synthesis, which is covered in the review by RADMAN and WAGNER (this volume). Some role of host-mediated DNA methylation is implemented in the regulation of gene expression of bacteriophage mu; this work is the subject of the review by KAHMANN (this volume).

We have been interested for some time in the methylation of DNA in *Bacillus subtilis*. This interest originated from studies on the mechanism of transformation and transfection. Here the use of restricting/modifying recipient cells was a convenient biological way to asses the biological nature and processing of incoming DNA (TRAUTNER et al. 1974; GÜNTHERT et al. 1977; BRON et al. 1980a, b; CANOSI et al. 1981). Later we realized that

---

* See also note added in proof, p. 173

1 Max-Planck-Institut für Molekulare Genetik, Ihnestraße 63–73, D-1000 Berlin 33 (Dahlem)

Current Topics in Microbiology and Immunology, Vol. 108
© Springer-Verlag Berlin·Heidelberg 1984

Table 1. Restriction and modification systems of *B. subtilis*

R/M system[a]	Strain origin[a]	Gene designation	Map positions[a,b] of *B. subtilis* 168	Sequence and modification specificity[c]	Isoschizomer
*Bsu*M	168	*hsr*M *hsm*M	13.0	CTĊGAG	*Xho*I
*Bsu*R	R	*hsr*R *hsm*R	16.4	GGĊC	*Hae*III
*Bsu*C	IAM1247	*hsr*C *hsm*C	16.0	n.d.	–
*Bsu*F	IAM1231	*hsr*F *hsm*F	(16.0)	ĊCGG	*Msp*I
*Bsu*E	IAM1231	*hsr*E *hsm*E	95.5	CGCG	*Tha*I
*Bsu*B	IAM1247	*hsr*B *hsm*B	97.0	CTGCAG	*Pst*I

[a] IKAWA et al. (1981)
[b] The coordinates for map positions are those given in the *B. subtilis* map of the Catalog of Strains of the *Bacillus* Genetic Stock Center (1982)
[c] References to target sequences: *Bsu*R: BRON et al. (1975); *Bsu*B: SHIBATA et al. (1979); *Bsu*M, *Bsu*E, *Bsu*F (JENTSCH, 1983). The methylated bases (Ċ) were determined by GÜNTHERT et al. (1978) for *Bsu*R and by JENTSCH (1983) for *Bsu*M and *Bsu*F. The target sequence for *Bsu*C has not been determined (n.d.)

many phages in *B. subtilis* have the potential of self-methylation (see Sect. 3), which offered convenient access to methyltransferase (MTase) genes and enzymes (TRAUTNER et al. 1980). Beyond that, sporulation in *B. subtilis*, which many researchers consider a process of differentiation, has been brought into connection with DNA methylation.

In this review we summarize information on DNA methylation systems encoded by *B. subtilis* and some of its bacteriophages. Since all MTases of *B. subtilis* described which are associated with restriction enzymes are representative of type-II modification enzymes, aspects of *B. subtilis* methylation systems comparative with those of other organisms are also covered.

2 Bacterial DNA Methylation[1]

2.1 Genetic Aspects

Analyzing restriction and modification of phages *Φ*105 and SPP1, six restriction and modification (R/M) systems could be identified in various *B. subtilis* strains (Table 1). Of these R/M systems, only one – *Bsu*M – was originally

1 Description of R/M-systems
  Genotypes: hsr/hsm: host-specific restriction/modification; psm: phage-specific methylation. *Phenotypes*: r⁺/r⁻ restriction proficient/deficient; m⁺/m⁻: modification proficient/deficient. The specificities of the bacterial R/M and the phage methylation systems are indicated by capital letters, indicating strain provenience following the genotype/phenotype description

Fig. 1. Genetic map of *B. subtilis* 168 (adapted from the map of the "*Bacillus* Genetic Stock Center"). The outer circle shows several physiological markers as reference. The inner circle shows the attachment sites of phages SPβ and φ3T/ρ11 and the various sites discussed in this paper. The attachment site of SPR is probably identical to that of SPβ (Anagnostopoulos, personal communication). The position of *hsr*F is not entered; it has tentatively been located at the position of *hsr*R and *hsr*C (IKAWA et al. 1981)

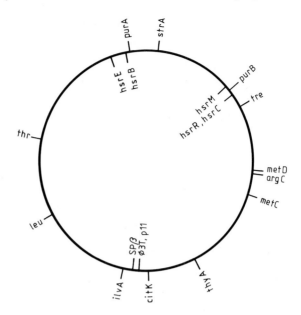

found in the commonly used Marburg strain, *B. subtilis* 168 (SHIBATA and ANDO 1974). All others were introduced through transformation from other *B. subtilis* strains into the 168-type strain (IKAWA et al. 1980). The chromosomal locations of genes specifying these R/M systems have been determined through transformation and transduction (TRAUTNER et al. 1974; IKAWA et al. 1981). These genes do not correspond to established prophage integration sites, nor have R/M systems been found to be encoded by plasmids. Whereas *hsr*M, *hsr*E, and *hsr*B map at different positions in the chromosome, *hsr*R, *hsr*C, and possibly *hsr*F are located at one and the same position (IKAWA et al. 1981) (Fig. 1). The *hsr*R, *hsr*C, and *hsr*F systems have never been found to coexist within one cell. This observation, together with the mapping experiments, has led to the supposition that *hsr*R, *hsr*C, and *hsr*F are allelic systems (IKAWA et al. 1980, 1981). No experiments have been performed to determine whether in transformation of the *B. subtilis* strain 168 the *hsr* genes from other strains substitute for homologous pseudogenes, or whether they are genes added to its chromosome.

The mapping experiments performed did not allow discrimination between the localization of the genes for the corresponding restriction and modification enzymes. However, these genes are separable from each other. This finding followed from the discovery of modification-proficient, but restriction-deficient mutants. Also, chromosomal DNA restriction fragments could be cloned under conditions which would select for methylation-proficient cells which exclusively contained the gene for an MTase (*Bsp*I, *Msp*I, *Bsu*RI) and not for the matching restriction enzyme (SZOLOMÁNYI et al. 1980; WALDER pers. communication; KISS and BALDAUF 1983). In spite of such genes being separable from each other, *hsr* and *hsm* genes appear to map close to each other. This was shown specifically for the *hsr*R and *hsm*R genes by A. Anagnostopoulos (unpublished results 1980). The same close proximity of R/M genes has been observed in the cloning of restriction genes of other organisms. Here the cloned DNA fragments expressed modifying as well as restricting activity of *Hha*II (MANN et al. 1978), *Eco*RI (GREENE et al. 1981; NEWMAN et al. 1981) and *Pst*I (WALDER et al. 1981).

2.2 Biochemical Characterization

The recognition sequences of *Bsu*B (CTGCAG, isoschizomer of *Pst*I) and of *Bsu*R (GGCC, isoschizomer of *Hae*III) have been known for some time (SHIBATA et al. 1979; BRON and MURRAY 1975). The identification of the target sequence of *Bsu*M as CTCGAG (*Xho*I), of *Bsu*E as CGCG (*Tha*I) and of *Bsu*F as CCGG (*Msp*I) has recently been achieved in our laboratory (JENTSCH 1983). The methylated base has been established in the *Bsu*M (*Xho*I) recognition site to be the central C in the sequence CTCGAG. In the recognition sequence CCGG of *Bsu*F it is the outer C of the sequence which is methylated. This followed from the sensitivity of DNA from strains with the *Bsu*F system to *Hpa*II degradation and its resistance to degradation with R·*Msp*I. The *Bsu*M MTase thus has the same specificity as that discovered by us earlier in strain Q of *B. subtilis* (JENTSCH et al. 1981). The recognition sequence of *Bsu*C has not yet been established (Table 1).

Among the *B. subtilis* cellular R/M systems both the restriction and modification enzymes of *Bsu*R have been characterized and purified to homogeneity (BRON and HÖRZ 1980; GÜNTHERT et al. 1981a). The restriction endonuclease produces a flush end cut between G and C of the cognate sequence GGCC (BRON and MURRAY 1975), whereas *Bsu*R modification is mediated by methylation of the central C in both strands (GÜNTHERT et al. 1978). While R·*Bsu*R requires only Mg^{++} and M·*Bsu*R only S-adenosyl-L-methionine (AdoMet), neither enzyme requires ATP for activity (BRON et al. 1975; GÜNTHERT et al. 1981b). The central C of GGCC is also the modification site of many other restriction enzymes such as *Bal*I (TGGCCA), *Xma*III (CGGCCG), or *Apa*I (GGGCCC) which have an internal GGCC in their recognition sequence (TRAUTNER, T.A., unpublished results 1982). The *Bsu*R MTase was characterized after 4000-fold purification (GÜNTHERT et al. 1981a, b). As with M·*Hpa*II (YOO and AGARWAL 1980), two very similar forms of enzyme were identified. The major class of enzyme had a mol.wt. of 42000, the minor class of enzyme (representing some 10% of the total activity) of 41000, with different net charge. Preliminary peptide mappings of the two forms revealed structural relatedness (GÜNTHERT U, unpublished 1982), which has also been described for M·*Hpa*II, suggesting one form of the enzyme to be the precursor of the other (YOO and AGARWAL 1980). Several properties of M·*Bsu*R may be related to those of other enzymes: The molecular weight of M·*Bsu*R is very similar to that of other highly purified type-II enzymes such as M·*Eco*RI: GAÅTTC (RUBIN and MODRICH 1977); M·*Hpa*II: CĊGG (YOO and AGARWAL 1980); M·*Tth*HB8I: TCGÅ (SATO et al. 1980). Like these enzymes, M·*Bsu*R is active in its monomeric form. Also, the reaction constant of

M·*Bsu*R with the substrates DNA ($K_m = 10^{-9}$ M) and AdoMet ($K_m = 10^{-7}$ M) are of the same order of magnitude as values observed for M· *Eco*RI and M·*Tth*HB8I (RUBIN and MODRICH 1977; SATO et al. 1980). However, the characteristics mentioned above do not seem to be typical for type-II modification enzymes at large; e.g., the active form of the M·*Pst*I enzyme is composed of four identical subunits with mol. wts. of 105000 each (LEVY and WELKER 1981).

Information on the mechanics of methylation by M·*Bsu*R were obtained by using different DNA substrates. We have observed that *Bsu*R-modified DNA was a competitive inhibitor of methylation with nonmodified DNA. Possibly, the M·*Bsu*R enzyme at first binds reversibly to DNA, irrespective of the presence of unmethylated recognition sites. Comparing two nonmodified DNAs as substrates, differing in the content of GGCC sites by a factor of 100, we realized that the rates of methylation were the same provided the molarity of GGCC sites of the two substrates was identical. Furthermore, the rate of methylation remained constant until the DNA was completely methylated (GÜNTHERT et al. 1981 b). Obviously, one methyl group is transferred per reaction step by the monomeric MTase, with each methylation occurring independently of the other. Hemimethylated sites are protected against double-strand cleavage by the R·*Bsu*R enzyme. M·*Bsu*R interacts with hemimethylated DNA as substrate for methylation at a velocity two times higher than with nonmethylated DNA. This is analogous to observations of methylation by the *dam*-MTase (HERMAN and MODRICH 1981), but also comparable to M·*Eco*RI (RUBIN and MODRICH 1977) which methylates nonmethylated and hemimethylated DNAs with equal efficiency. With respect to their kinetic behavior, these enzymes are therefore basically different from type-I MTases, which methylate hemimethylated DNA some 100 times more efficiently than they do nonmethylated DNA (VOVIS et al. 1974).

2.3 Some Physiological Considerations of Methylation

The obvious role of the MTases discussed is to provide modification. It is not clear what role DNA methylation plays in *B. subtilis* in addition to modification. The extent of DNA methylation appears to have some correlation with the physiological state of *B. subtilis* cells; for example, GANESAN (1979) has observed that cells grown under starvation conditions, which induce competence and prophage (YASBIN et al. 1975), have elevated levels of MTase activity. Neither the specificity nor the genetic determinants which are responsible for this methylation have been determined. GÜNTHERT (1975) observed a marked increase in the extent of DNA C methylation of *B. subtilis*, *B. megatherium*, and *B. brevis* during the growth from vegetative cells to sporulation. This has led to speculations about the involvement of DNA methylation in the regulation of sporulation.

No equivalent of a methylation-controlled proofreading system seems to be operating in *B. subtilis*. This was concluded from the absence in *B. subtilis* DNA of methylation comparable to that provided by the *dam/dcm* system in *E. coli* DNA. Specifically with pHV15, a shuttle plasmid for

E. coli, Staphylococcus aureus, and *B. subtilis,* DREISEIKELMANN and
WACKERNAGEL (1981) demonstrated that neither *S. aureus* nor *B. subtilis*
have MTases with *dam* or *dcm* specificity.

The regulation of expression of an R/M system was studied in some
detail with the *Bsu*M system, where a close connection between the sporulation proficiency and the expression of *Bsu*M restriction became obvious.
FUČIK et al. (1982) observed that sporulation deficiency caused by an early
mutation in the sporulation pathway (*spo*OA) abolished *Bsu*M restriction,
leaving the *Bsu*M modification unaffected. This suggested that the *hsr*M
and *hsm*M genes are not coordinately regulated. The absence of expression
of the *hsr*M gene in *spo*OA cells could be reversed if such cells became
resistant to streptomycin. It is hoped that an elucidation of the connection
between sporulation proficiency and expression of the *Bsu*M system will
benefit from the recent identification of the biochemical specificity of the
*Bsu*M system (JENTSCH 1983). No information is available about the regulation of expression of other MTase genes in *B. subtilis.*

3 Bacteriophage DNA Methylation

DNAs of bacteriophages entering host cells are substrates for the cellular
restriction endonucleases and MTases. In an encounter of a nonmodified
phage DNA with a restriction/modifying cell, the establishment of a replicating phage genome depends, on the one hand, on the relative activities with
which the two types of enzymes react with the phage DNA. On the other
hand, bacteriophages, including those of *B. subtilis,* have developed several
mechanisms which enable them to actively overcome the barrier set up
by the R/M system of infected cells.

Several *B. subtilis* phages contain unusual bases in their DNAs which
make them less sensitive to restriction (HEMPHILL and WHITELEY 1975; ITO
et al. 1975).

Antirestriction activity was identified in phages φNR2 (MAKINO et al.
1979) and φ1 (MAKINO and SAITO 1980), and was suggested to be operative
in SPO1 as well (REEVE et al. 1980). Interestingly, infection of *Bsu*R r$^+$m$^+$
cells with SPO1 also seemed to inactivate the prevailing modification enzyme, since SPO1 progeny from such infected cells was without *Bsu*R modification (GÜNTHERT et al. 1975; REEVE et al. 1980). A heterologous antirestriction activity affecting the restriction of SP10 by the *Bsu*M system was
elicited by bacteriophage SP18, when SP18 infection preceded that of SP10
(WITMER and FRANKS 1981). In neither case is the biochemical basis of
antirestriction known.

Absence or paucity of cognate sequences of certain R/M systems in
phage DNAs has been recognized in various *B. subtilis* phages. This is
most significant with the recognition sequence GGCC, which is absent in
φ29 (ITO and ROBERTS 1979), and only four *Bsu*R sites rather than the
expected ∼300 are present in the DNA of phage SPO1 (REEVE et al. 1980).

In this review we discuss those aspects of bacteriophage resistance which
are a consequence of self-methylation. In these cases, the phage DNAs

Table 2. Methylation systems of *B. subtilis* bacteriophages

Designation of methylation system	Bacteriophage origin	Gene designation[a]	Sequence and methylation specificity	Resistance to restriction with
M·SPRI	SPR	*psm*SPR	GGC̊C	*Bsu*R (*Hae*III)
M·SPRII			C̊C̊GG	*Hpa*II and *Bsu*F (*Msp*I)
M·SP*β*I	SP*β*	*psm*SP*β*	GGC̊C	*Bsu*R (*Hae*III)
M·SP*β*II			GC̊NGC	*Fnu*4HI
M·SP*φ*3TI	*φ*3T	*psmφ*3T	GGC̊C	*Bsu*R (*Hae*III)
M·SP*φ*3TII			GC̊NGC	*Fnu*4HI
M·SP*ρ*11I	*ρ*11	*psmρ*11	GGC̊C	*Bsu*R (*Hae*III)
M·SP*ρ*11II			GC̊NGC	*Fnu*4HI

[a] The gene designation assumes that in all phages *one* gene determines all methylation activities observed

do contain restriction sequences recognizable by host cell restriction systems. These sites are obligatorily modified, however, due to the presence in phage genomes of genes specifying modification-type MTases. These genes are expressed early in phage development. In contrast to host R/M systems, the *B. subtilis* phage-borne modification enzymes have never been found in association with restriction enzymes.

3.1 Genetic Aspects

Self-methylation leading to resistance against R·*Bsu*R restriction was observed in the temperate *B. subtilis* phages SPR, SP*β*, *φ*3T, and *ρ*11. The methylation capacity of these phages was established in an analysis of the sensitivity of their DNAs to a number of restriction enzymes. All phage DNAs were nondegradable with R·*Bsu*R and its isoschizomer R·*Hae*III (NOYER-WEIDNER et al. 1983). The DNA of SPR was also found to be resistant to R·*Hpa*II and R·*Msp*I treatment; DNAs of SP*β*, *φ*3T, and *ρ*11 were resistant to R·*Fnu*4HI degradation (Table 2).

Evidence in favor of the interpretation that the observed MTase activity was phage determined came from genetic studies. Phage mutants of SPR and *φ*3T deficient in *Bsu*R-methylating activity were selected on the basis of their inability to grow on *Bsu*R-restricting strains. DNAs of such mutants were sensitive to *Bsu*R restriction, and their 5-methylcytosine (5 mC) content was significantly reduced (NOYER-WEIDNER et al. 1983). Similarly, phage mutants of SPR defective in *Msp*I/*Hpa*II-methylating activity were selected using *B. subtilis* strains with the *Bsu*F R/M system. Interestingly, the majority of mutants found had lost not only their *Bsu*R-methylating potential, but also the capacity to express the nonselected methylation activities M·SPRII and M·*φ*3TII (NOYER-WEIDNER et al. 1983). Only 20% of the selected M·SPRI- and M·*φ*3TI-deficient mutants had retained the capacity to perform the nonselected types of methylation. Crosses between such a "single" mutant and a mutant in which both enzymatic deficiencies were affected yielded wild-type (WT) recombinants which were proficient in all methylation activities (NOYER-WEIDNER et al. 1983). A variety of interpretations could accommodate these findings. Cloning experiments and enzyme purification to be reported later strongly suggested that one gene, and hence one enzyme, was responsible for both methylation activities. A mutation causing loss of both MTase activities might be in a location which controls a general step in the MTase reaction, whereas a region determining specificity of recognition might be the target of the "single" mutation.

The *psm* genes of phages SPR, SPβ, φ3T, and ρ11 are interchangeable and the latter three genes are structurally related. This was first observed in heterologous phage crosses. Methylation proficient phages are generated, for example, when a *B. subtilis* cell, lysogenized with an MTase-deficient mutant of SPR is transfected with intact DNA of a restriction endonuclease-generated digest of φ3T WT DNA, neither of which alone is active in the transfection of a nonlysogenic cell (NOYER-WEIDNER et al. 1983). The recombinant phages obtained in this cross or in analogous crosses between other phages always had the methylation potential of the donor phage DNA (NOYER-WEIDNER et al. 1983). Obviously, the *psm* gene of the donor DNA substitutes for the defective MTase gene of the prophage.

Further indication for the homology between various phage *psm* genes came from DNA/DNA hybridization experiments (NOYER-WEIDNER et al. 1983), in which a labeled DNA fragment containing only the *psm*SPR gene was used as a hybridization probe against digests of the DNAs of the phages concerned. The probe hybridized to only one restriction fragment of each phage DNA. In all cases it was that fragment which had independently been shown to contain the phage's *psm* gene. Hybridization was also observed against a restriction fragment of the related bacteriophage Z. This phage does not have the capacity for self-methylation. To interpret the presence of homology to the *psm* SPR gene, we assume that the chromosome of Z contains a *psm* "pseudo gene". Interestingly, no hybridization of phage-borne *psm* genes was observed with *hsm*R-proficient bacterial DNA. Therefore, the bacterial- and phage-specified MTase genes which code for MTases with identical specificity appear to derive from different ancestral genes. A comparison of the organization of the enzymes encoded by *B. subtilis* and the phages discussed here will be of interest for defining the parameters which describe MTases with identical specificity.

3.2 Cloning Experiments

In an effort to characterize the genomic organization of MTase genes as well we have cloned DNA fragments obtained in restriction enzyme digests of SPR and φ3T DNA, which contain the phage MTase gene. From SPR DNA a 4.5-kb *EcoRI* fragment with the MTase gene could readily be inserted into the only *EcoRI* site of a λ-phage insertion vector (BEHRENS et al. 1983). Expression of SPR-specific MTase activities was observed when the DNA of such hybrid λ/SPR phages was subjected to degradation with R · *BsuR*, R · *MspI*, and R · *HpaII* (MONTENEGRO et al. 1983). Expression was not complete; i.e., digests with the three restriction enzymes indicated the presence of significant amounts of DNA which had not become methylated. This expression was under the control of the λ P_{RM} promoter, since expression was observed in only one of the two orientations of the integrated SPR fragment in λ (MONTENEGRO et al. 1983). Hybrid λ/SPR phages containing the corresponding DNA fragment derived from the M·SPRI-deficient mutants showed the same methylation potential as that of the original mutants. The λ/SPR hybrid phages carrying SPR-specific methylation were not distinguishable in their biological properties from nonmethylated λ phages; obviously the foreign methylation of λ DNA, imposed through the insertion of the *psm* SPR gene, is without any influence on λ gene expression and DNA replication.

Within the cloned 4.5-kb fragment of SPR DNA we were subsequently able to identify a region of 2 kb on which the mutations in the *psm* SPR gene were located. This region was contained in a 2.3-kb *Pst*I fragment, which – when cloned onto a *B. subtilis* phage vector – was also expressed in *B. subtilis* (BEHRENS et al. 1983). With this *B. subtilis* phage containing an SPR fragment, both the M·SPRI and the M·SPRII methylations were expressed when the SPR DNA was derived from SPR WT. Only M·SPRII activity or no activity at all could be detected when the SPR DNA was derived from either one of the mutant phages observed. From these cloning results we conclude that the 2-kb fragment of SPR DNA contains the structural gene for an enzyme with both MTase activities described. Analogous results were obtained with the DNA containing an MTase gene derived from phage ϕ3T (NOYER-WEIDNER M., unpublished results 1983).

Recently KISS and BALDAUF (1983) have reported on the cloning of the same *Eco*RI fragment of SPR containing the *psm* SPR gene in pBR322. In this vector as well, both the M·SPRI and the M·SPRII activities were expressed.

3.3 Biochemical Characterization

Purification of SPR- and ϕ3T-specified MTases has been in progress for some time (Günthert, unpublished results). During several chromatographic steps the MTase activities designated as I and II in Table 2 did not separate from each other. This result is in agreement with our proposition that one and the same enzyme is responsible for both types of methylation. Furthermore, from these experiments it was learned that the SPR- and ϕ3T-specified MTases have the characteristics of type-II modification enzymes. They are active in their monomeric forms and they require AdoMet, but not Mg^{++} and ATP as cofactors. The molecular weight of M·SPR was tentatively determined to be 47000. This molecular weight corresponds quite well to the coding capacity of the cloned SPR DNA containing the *psm* SPR gene.

3.4 Biological Considerations

The obvious biological role of the phage-specified MTases is to permit the phages to overcome barriers set up by the R/M systems of infected cells. In this way one can rationalize the presence of those phage methylation systems which correspond to host-specific R/M systems, such as *Bsu*R or *Bsu*F. It will be interesting to see whether other host R/M systems which could be overcome by phages having the potential for *Fnu*4HI- or *Hpa*II-specific modification can be identified in *B. subtilis*.

We have no indication that phage-mediated methylation plays a role additional to the one described. Apparently the capacity for self-methylation is not an essential function, since the MTase-deficient mutants of SPR and φ3T have no selective disadvantage in comparison with the WT phages as long as growth takes place in nonrestricting hosts. However, the *psm* gene of SP*β* was found to be highly conserved, even in the absence of selective pressure for its maintenance: prophage SP*β*, which had been present for an indeterminable number of generations in a mutator strain of *B. subtilis* (r⁻m⁻) still had an unaltered methylation potential (TRAUTNER T.A., unpublished results 1980).

Regarding the regulation of expression of *psm* genes, we have seen that they are expressed early in phage development. The DNAs of cells lysogenic for such phages can be degraded by restriction enzymes corresponding to the phage-coded MTases. This indicates that the prophage genes are not expressed, at least in the majority of cells.

4 Concluding Remarks

All MTases identified in *B. subtilis* (Table 1) have been found in association with corresponding restriction endonucleases. They methylate C within their respective cognate sequences. No function of these enzymes other than to provide modification was established. Genes determining R/M map at various locations within the *B. subtilis* chromosome (Fig. 1). All R/M systems, with the exception of *Bsu*M, were introduced into *B. subtilis* strain 168 from other *B. subtilis* strains.

A group of large temperate *B. subtilis* bacteriophages encode their own MTases (Table 2). These enzymes have dual specificity; they methylate the central C of the recognition sequence GGCC, and, in addition, either the neighboring Cs in the sequence CCGG (SPR) or the central C of GCNGC (SP*β*, φ3T, *p*11). By means of DNA/DNA hybridization techniques, homology was observed among some MTase genes of the phages. No homology could be discovered between these phage genes and *B. subtilis* chromosomal DNA.

Acknowledgments. We are grateful to S. JENTSCH and M. NOYER-WEIDNER for their helpful comments and discussions during the preparation of this review and for providing data prior to publication. Experimental work from the authors' laboratory was supported in part by the *Deutsche Forschungsgemeinschaft* (Tr 25/9-3).

References

Behrens B, Pawlek B, Morelli G, Trautner TA (1983) Restriction and modification in *Bacillus subtilis*: Construction of hybrid *λ* and SPP1 phages containing a DNA methyltransferase gene from *B. subtilis* phage SPR. MGG 189:10–16

Bron S, Murray K (1975) Restriction and modification in *B. subtilis*. Nucleotide sequence recognized by restriction endonuclease R · *Bsu*R from strain R. MGG 143:25–33

Bron S, Hörz W (1980) Purification and properties of the *Bsu* endonuclease. Methods Enzymol 65:112–132

Bron S, Murray K, Trautner TA (1975) Restriction and modification in *B. subtilis*. Purification and general properties of a restriction endonuclease from strain R. MGG 143:13–23

Bron S, Luxen E, Trautner TA (1980a) Restriction and modification in *B. subtilis*. The rate of homology between donor and recipient DNA in transformation and transfection. MGG 179:111–117

Bron S, Luxen E, Venema G, Trautner TA (1980b) Restriction and modification in *B. subtilis*. Effects on transformation and transfection with native and single-stranded DNA. MGG 179:103–110

Canosi U, Lüder G, Trautner TA, Bron S (1981) Restriction and modification in *B. subtilis*: Effects on plasmid transformation. In: Polsinelli M, Mazza G (eds) Transformation 1980, Proceedings of the Fifth European Meeting on Bacterial Transformation and Transfection, Florence. Cotswold, Oxford, pp 179–187

Dreiseikelmann B, Wackernagel W (1981) Absence in *Bacillus subtilis* and *Staphylococcus aureus* of the sequence-specific deoxyribonucleic acid methylation that is conferred in *Escherichia coli* K-12 by the *dam* and *dcm* enzymes. J Bacteriol 147:259–261

Fučik V, Grünnerová H, Zadražil S (1982) Restriction and modification in *Bacillus subtilis* 168. Regulation of *hsr*M (nonB) expression in *spo*OA mutants and effects on permissiveness for φ15 and φ105 phages. MGG 186:118–121

Ganesan AT (1979) Genetic recombination during transformation in *Bacillus subtilis*: appearance of a deoxyribonucleic acid methylase. J Bacteriol 139:270–279

Greene PJ, Gupta M, Boyer HW, Brown WE, Rosenberg JM (1981) Sequence analysis of the DNA encoding the *Eco*RI endonuclease and methylase. J Biol Chem 256:2143–2153

Günthert U (1975) DNA Methylierung während der Entwicklung von Pro- und Eukaryonten. PhD Thesis, Univ. Tübingen

Günthert U, Stutz J, Klotz G (1975) Restriction and modification in *Bacillus subtilis*. The biochemical basis of modification against Endo R · *Bsu*R restriction. MGG 142:185–191

Günthert U, Pawlek B, Stutz J, Trautner TA (1976) Restriction and modification in *Bacillus subtilis*: Inducibility of a DNA-methylating activity in nonmodifying cells. J Virol 20:188–195

Günthert U, Pawlek B, Trautner TA (1977) Transfection as a tool in the study of modification. In: Portolés A, Lopez R, Espinoza M (eds) Modern trends in bacterial transformation and transfection. Elsevier, Amsterdam, pp 249–256

Günthert U, Storm K, Bald R (1978) Restriction and modification in *Bacillus subtilis*. Localization of the methylated nucleotide in the *Bsu*RI recognition sequence. Eur J Biochem 90:581–583

Günthert U, Freund M, Trautner TA (1981a) Restriction and modification in *Bacillus subtilis*: two DNA methyltransferases with *Bsu*RI specificity. I. Purification and physical properties. J Biol Chem 256:9340–9345

Günthert U, Jentsch S, Freund M (1981b) Restriction and modification in *Bacillus subtilis*: two DNA methyltransferases with *Bsu*RI specificity. II. Catalytic properties, substrate specificity, and mode of action. J Biol Chem 256:9346–9351

Hemphill HE, Whiteley HR (1975) Bacteriophages of *Bacillus subtilis*. Bacteriol Rev 39:257–315

Herman GE, Modrich P (1982) *Escherichia coli dam* methylase. J Biol Chem 257:2605–2612

Ikawa S, Shibata T, Ando T, Saito H (1980) Genetic studies on site-specific endodeoxyribonucleases in *Bacillus subtilis*: multiple modification and restriction systems in transformants of *B. subtilis* 168. MGG 177:359–368

Ikawa S, Shibata T, Matsumoto K, Iijima T, Saito H, Ando T (1981) Chromosomal loci of genes controlling site-specific restriction endonuclease of *B. subtilis*. MGG 183:1–6

Ito J, Roberts RY (1979) Unusual base sequence arrangement in phage φ29 DNA. Gene 5:1–7

Ito J, Kawamura F, Duffy JJ (1975) Susceptibility of nonthymine-containing DNA to four bacterial restriction endonucleases. FEBS Lett 55:278–281

Jentsch S, Günthert U, Trautner TA (1981) DNA methyltransferases affecting the sequence 5'CCGG. Nucleic Acids Res 9:2753–2759

Jentsch S (1983) Restriction and Modification in *Bacillus subtilis*: The sequence specificity of restriction/modification systems *Bsu*M, *Bsu*F, and *Bsu*E. J. Bact. (in press)

Kiss A, Baldauf F (1983) Molecular cloning and expression in *Escherichia coli* of two modification methylase genes of *Bacillus subtilis*. Gene 21:111–119

Levy WP, Welker NE (1981) DNA modification methylase from *Bacillus stearothermophilus*. Biochemistry 20:1120–1127

Makino O, Saito H (1980) *Bacillus subtilis* phage ϕ1 overcomes host-controlled restriction by producing *Bam*Nx inhibitor protein. MGG 179:463–468

Makino O, Kawamura F, Saito H, Ikeda Y (1979) Inactivation of restriction endonuclease *Bam*Nx after infection with phage ϕNR2. Nature 277:64–66

Mann MB, Rao RN, Smith HO (1978) Cloning of restriction and modification genes in *E. coli*: The *Hha*II system from *Haemophilus haemolyticus*. Gene 3:97–112

Montenegro MA, Pawlek B, Behrens B, Trautner TA (1983) Restriction and modification in *Bacillus subtilis*: Expression of the cloned methyltransferase gene from *B. subtilis* phage SPR in *E. coli* and *B. subtilis*. MGG 189:17–20

Newman AK, Rubin RA, Kim S-H, Modrich P (1981) DNA sequences of structural genes for *Eco*RI DNA restriction and modification enzymes. J Biol Chem 256:2131–2139

Noyer-Weidner M, Pawlek B, Jentsch S, Günthert U, Trautner TA (1981) Restriction and modification in *Bacillus subtilis*: Gene coding for a *Bsu*R-specific modification methyltransferase in the temperate bacteriophage ϕ3T. J Virol 38:1077–1080

Noyer-Weidner M, Jentsch S, Pawlek B, Günthert U, Trautner TA (1983) Restriction and modification in *Bacillus subtilis*: The DNA methylation potential of the related bacteriophages Z, SPR, SPβ, ϕ3T and ρ11. J Virol 46:446–453

Reeve JN, Amann E, Tailor R, Günthert U, Scholz K, Trautner TA (1980) Unusual behaviour of SPO1 DNA with respect to restriction and modification enzymes recognizing the sequence 5'GGCC. MGG 178:229–231

Rubin RA, Modrich P (1977) *Eco*RI methylase. Physical and catalytic properties of the homogeneous enzyme. J Biol Chem 252:7265–7272

Sato S, Nakazawa K, Shinomiya T (1980) A DNA methylase from *Thermus thermophilus* HB8. J Biochem 88:737–747

Shibata T, Ando T (1974) Host-controlled modification and restriction in *Bacillus subtilis*. MGG 131:275–280

Shibata T, Ikawa S, Komatsu Y, Ando T, Saito H (1979) Introduction of host-controlled modification and restriction systems of *Bacillus subtilis* IAM 1247 into *Bacillus subtilis* 168. J Bacteriol 139:308–310

Szomolányi É, Kiss A, Venetianer P (1980) Cloning the modification methylase gene of *Bacillus sphaericus* R in *Escherichia coli*. Gene 10:219–225

Trautner TA, Pawlek B, Bron S, Anagnostopoulos C (1974) Restriction and modification in *B. subtilis*. Biological aspects. MGG 131:181–191

Trautner TA, Pawlek B, Günthert U, Canosi U, Jentsch S, Freund M (1980) Restriction and modification in *Bacillus subtilis*: Identification of a gene in the temperate phage SPβ coding for a *Bsu*R specific modification methyltransferase. MGG 180:361–367

Walder RY, Hartley JL, Donelson JE, Walder JA (1981) Cloning and expression of the *Pst*I restriction-modification system in *Escherichia coli*. Proc Natl Acad Sci USA 78:1503–1507

Witmer H, Franks M (1981) Restriction and modification of bacteriophage SP10 DNA by *Bacillus subtilis* Marburg 168: Stabilization of SP10 DNA in restricting hosts preinfected with a heterologous phage, SP18. J Virol 37:148–155

Vovis GF, Horiuchi K, Zinder ND (1974) Kinetics of methylation of DNA by a restriction endonuclease from *Escherichia coli* B. Proc Natl Acad Sci USA 71:3810–3813

Yasbin RE, Wilson GA, Young FE (1975) Transformation and transfection in lysogenic strains of *Bacillus subtilis*: Evidence for selective induction of prophage in competent cells. J Bacteriol 121:296–304

Yoo OJ, Agarwal KL (1980) Isolation and characterization of two proteins possessing *Hpa*II methylase activity. J Biol Chem 255:6445–6449

Effects of DNA Methylation on Mismatch Repair, Mutagenesis, and Recombination in *Escherichia coli*

M. RADMAN and R. WAGNER

1 Introduction

In the DNA of *Escherichia coli* K12 approximately 2% of the adenines and 1% of the cytosines are methylated to form 6-methyladenine (6-mA) and 5-methylcytosine (5-mC) respectively (reviews, HATTMAN 1981; MARINUS 1982). Three different types of *E. coli* mutants deficient in DNA methylation have been isolated and characterized: (a) *dam* mutants, which are deficient in adenine methylation of GATC sequences (MARINUS and MORRIS 1973; LACKS and GREENBERG 1977; HATTMAN et al. 1978); (b) *dcm* (or *mec*) mutants, which are deficient in cytosine methylation of CC(A/T)GG sequences (MARINUS and MORRIS 1973; HATTMAN et al. 1973); and (c) *hsd* mutants, which are deficient in adenine methylation of AAC (N6) GTGC sequences (review, YUAN 1981).

Hsd controlled methylation protects the cell's DNA from restriction by its own restriction endonuclease, *Eco* K12. Therefore, *hsd* mutants are viable only if the cell is also restriction-deficient. *Dcm* mutants have no easily discernible phenotype, however it has been demonstrated that they have lost some mutational hot spots. *Dam* mutations are highly pleiotropic. Relative to wild-type cells, *dam* mutants (a) have increased spontaneous mutation rates (mutator phenotype) (MARINUS and MORRIS 1975), (b) show increased levels of recombination (hyper-rec phenotype) (MARINUS and KONRAD 1976), and (c) are more sensitive to base analogs, EMS, MNNG, MMS, and UV (GLICKMAN et al. 1978; JONES and WAGNER 1981; MARINUS and MORRIS 1974). In addition, *dam rec A*, *dam rec B*, *dam lex A* and *dam pol A* double mutants are inviable (MARINUS and MORRIS 1975). Several of the phenotypes of *dam* mutants, e.g., sensitivity to mutagenic agents

Institut Jacques Monod, C.N.R.S., Université Paris VII, Tour 43, 2 Place Jussieu, F-75251 Paris CEDEX

Current Topics in Microbiology and Immunology, Vol. 108
© Springer-Verlag Berlin·Heidelberg 1984

and the inviability of *dam rec* A double mutants, are abolished by *mut H*, *mut L*, *mut S*, *mut U* (*uvr E*), and *sin* mutations (GLICKMAN and RADMAN 1980; McGRAW and MARINUS 1980; MARINUS 1980; JONES and WAGNER 1981).

2 DNA Methylation and Mismatch Repair

The methylation of adenine in the sequence GATC has been shown to lag somewhat behind replication, such that GATC sequences in newly synthesized DNA strands are transiently unmethylated (MARINUS 1976). Thus, in the region immediately behind the replication fork, newly synthesized strands can be distinguished from parental strands by their degree of methylation. It was suggested that the *E. coli* mismatch repair system might act in such regions to remove replication errors from newly synthesized strands (WAGNER and MESELSON 1976). This hypothesis was tested by annealing unmethylated strands of bacteriophage λ DNA (obtained from phages grown in *dam dcm* bacteria) to methylated strands of different genotype (obtained from phages grown in wild-type bacteria (RADMAN et al. 1980; MESELSON et al. 1980). The heteroduplexes thus formed contained one methylated and one unmethylated strand and base pair mismatches at sites of genetic differences between the two original phages. The heteroduplexes were introduced into *E. coli* cells by means of transfection and the phages produced were analyzed to determine genotype frequencies. It was found that the genotypes carried on the unmethylated strands were underrepresented among the progeny, suggesting that mismatch repair operates preferentially on unmethylated strands. However, it might have been that unmethylated strands were simply lost rather than repaired, i.e., they may have failed to produce progeny in sufficient numbers to be recovered. The occurrence of methyl-directed strand loss in this system has been excluded by the following results:

1. A nonrepairable mismatch shows no methyl-directed strand loss (MESELSON et al. 1980).
2. The methyl-directed bias in genotype output is abolished in mismatch-repair-deficient mutants (*mut L, mut S, mut H,* and *mut U*) (RADMAN et al. 1980, 1981).
3. Analysis of the genotypes of phages obtained from individual transfected cells in which one of a pair of mismatch markers was not repaired (i.e., both mutant and wild-type alleles of that marker were recovered from a single heteroduplex) showed methyl direction of repair at the other marker (C. Dohet, R. Wagner, and M. Radman, unpublished work).

It was also found that heteroduplexes in which the unmethylated strand was prepared from the DNA of phages grown in *dam dcm*$^+$ bacteria showed methyl direction of mismatch repair (unpublished experiments from the laboratories of M. Meselson and M. Radman). Thus it appears that adenine methylation of GATC sequences is responsible for the strand discrimination ability of the *E. coli* mismatch repair system. Presumably, the mutator phenotype of *dam* mutants is a result of undirected mismatch repair (theoretically, undirected mismatch repair has the same effect on mutation frequencies as no mismatch repair).

Since mismatch repair does not appear to operate on fully methylated DNA (MESELSON et al. 1980; PUKKILA et al. 1983), the speed with which methylation occurs following replication should determine the amount of time the mismatch repair system has to recognize and remove replication errors. *Dam* methylase overproducer strains, in which the *dam* gene is cloned in a multicopy plasmid, are mutators (HERMAN and MODRICH 1981). It may be that methylation occurs more rapidly in these strains than in wild-type strains, thus shortening the time available for mismatch repair. *Pur B ts* mutants, which contain low levels of the *S*-adenosylmethionine, required for *dam*-directed methylation are antimutators (GEIGER and SPEYER 1977), and the antimutator activity appears to depend on functional mismatch repair genes (*mut S* and *mut L*) (P. Schendel, personal communication). It may be that methylation is delayed in these strains, thus lengthening the time available for mismatch repair. These results suggest that no net selective pressure is operating on *E. coli* to decrease the spontaneous mutation rate by enhancing the efficiency of mismatch repair. In other words, it appears that either the benefit of the present mutation rate or the cost of reducing it may be too great to allow a significant decrease without creating a significant selective disadvantage for the cell.

The inviability of *dam rec A* double mutants can be overcome by the addition of a mutation affecting mismatch repair (*mut H*, *mut L*, or *mut S*) (GLICKMAN and RADMAN 1980; MC GRAW and MARINUS 1980). This suggests that some action of the mismatch repair system is responsible for the killing of the double mutant. It may be that the killing is caused by double-strand breaks in the DNA that cannot be repaired in *rec A* bacteria (the repair of double-strand breaks has been shown to be an SOS-inducible function, i.e., one that requires *rec A* and *lex A* products, KRASIN and HUTCHINSON 1981). Double-strand breaks could be produced by the mismatch repair system operating on *dam* (unmethylated) DNA either as result of simultaneous attack on both strands at a single mismatch or as a result of the overlap of excision tracts from repair events initiating on opposite strands at two relatively close mismatches. Whatever its mechanism, if this "mismatch-stimulated suicide" does occur, it could account for the findings that: (a) no SOS-induced mutator effect is observed in unirradiated λ phages replicating in UV-irradiated *dam* bacteria (P. CAILLET-FAUQUET, G. MAEN-HAUT-MICHEL, and M. RADMAN, unpublished work); and (b) *dam* mutants have a lower spontaneous mutation frequency than *mut H*, *mut L*, *mut S*, or *mut U* single mutants or *dam mut* double mutants (GLICKMAN and RAD-MAN 1980). Double-strand breaks have not been positively identified in *dam* DNA; however, if they occur as hypothesized above, they could also be the lesions responsible for the marker-independent hyper-rec phenotype of *dam* mutants.

3   Effects of Cytosine Methylation on Mutagenesis and Recombination

Although little is known regarding the role of cytosine methylation in *E. coli*, 5-mC residues have been found to be preferred sites for $C \rightarrow T$ transition mutation (COULONDRE et al. 1978). Under physiological conditions, cytosine deamination to form uracil is a common occurrence (LINDAHL and NYBERG 1974); the lesion is efficiently removed by the ura-DNA-glycosylase (review, LINDAHL 1982). However, deamination of 5-mC produces thymine, which is not a substrate for the glycosylase and which creates a G/T mixmatch. As expected, *dcm* (or *mec*) mutants (and *E. coli* B which are naturally *dcm*) do not show mutational hot spots at CC(A/T)GG sequences (COULONDRE et al. 1978). In *ung* mutants (deficient in ura-DNA-glycosylase) all cytosines show comparable frequencies of $C \rightarrow T$ transition (DUNCAN and MILLER 1980).

It has been found that hemimethylated DNA (i.e., DNA containing 5-mC in one strand but not in the other) is hyper-recombinogenic compared to DNA that is either fully methylated or fully unmethylated (KORBA and HAYS 1982). Mutants partially deficient in cytosine methylation (*arl*) exhibit increased recombination frequencies. In addition, λ phages grown in *arl* mutants show increased recombination frequencies when crossed in *arl*[+] bacteria. A *dcm* mutation abolishes the *arl* phenotype (KORBA and HAYS 1982). The mechanism of this stimulation of recombination by hemimethylated DNA is not known. (It may be that sister chromatid exchanges represent an analogous effect of cytosine hemimethylation in eukaryotes.)

The results of a recent study indicate that at least one of the two possible mismatches created by strand exchange between an amber mutant (TAG; created by 5-mC $\rightarrow$ T transition) and the methylated wild-type parent is repaired at high frequency (LIEB, 1983). This efficient action of the mismatch repair system could explain the high levels of recombination observed with this particular class of amber mutants (i.e., C*TA*GG) in standard genetic crosses (LIEB, to be published).

4   Summary

Escherichia coli K12 DNA contains methylated adenine (6-mA) and cytosine (5-mC) residues. Adenine methylation controlled by *hsd* is responsible for protection of the cell's DNA from restriction by its own restriction endonuclease. The *dam*-directed methylation of adenine in GATC sequences appears to direct the action of the mismatch repair system of *E. coli* in such a way that errors arising in newly synthesized (transiently unmethylated) DNA strands are corrected to the sequence of the parental strand. The precise mechanism of this strand direction is not known.

The extent of hemimethylated DNA, i.e., regions with one strand methylated and one strand unmethylated, appears (a) with respect to *dam*-directed adenine methylation, to influence the spontaneous mutation rate and (b) with respect to cytosine methylation, to influence the level of recombination.

Acknowledgments. Dr. MIROSLAV RADMAN was supported by a visiting professorship from the Université de Paris-Sud, Laboratoire de Génétique, Orsay, Dr. ROBERT WAGNER holds a fellowship from the Ligue Nationale Française Contre le Cancer.

References

Coulondre C, Miller JH, Farabaugh PJ, Gilbert W (1978) Molecular basis of base substitution hot spots in *Escherichia coli*. Nature 274:775–780

Duncan BK, Miller JH (1980) Mutagenic deamination of cytosine residues in DNA. Nature 287:560–561

Geiger JR, Speyer JF (1977) A conditional antimutator in *E. coli*. Mol Gen Genet 153:87–97

Glickman BW, Radman M (1980) *Escherichia coli* mutator mutants deficient in methylation-instructed mismatch correction. Proc Natl Acad Sci USA 77:1063–1067

Glickman BW, van der Elsen P, Radman M (1978) Induced mutagenesis in *dam⁻* mutants of *Escherichia coli*: a role for 6-methyladenine in mutation avoidance. Mol Gen Genet 163:307–312

Hattman S (1981) DNA methylation. The Enzymes 14:517–548

Hattman S, Schlogman S, Cousens L (1973) Isolation of a mutant of *Escherichia coli* defective in cytosine-specific deoxyribonucleic acid methylation activity and in partial protection of bacteriophage against restriction by cells containing the N-3 drug-resistance factor. J Bacteriol 115:1103–1107

Hattman S, Keister T, Gotteherr A (1978) Sequence specificity of DNA methylases from *Bacillus amyboliquefaciens* and *Bacillus brevis*. J Mol Biol 124:701–711

Herman GE, Modrich P (1981) *Escherichia coli* K12 clones that overproduce *dam* methylase are hypermutable. J Bacteriol 145:644–646

Jones M, Wagner R (1981) *N*-Methyl-*N'*-nitro-*N*-nitrosoguanidine sensitivity of *E. coli* mutants deficient in DNA methylation and mismatch repair. Mol Gen Genet 184:562–563

Korba BE, Hays JB (1982) Partially deficient methylation of cytosine in DNA at CCA/TGG sites stimulates genetic recombination of bacteriophage lambda. Cell 28:531–541

Krasin F, Hutchinson F (1981) Repair of DNA double-strand breaks in *Escherichia coli* cells requires synthesis of proteins that can be induced by UV light. Proc Natl Acad Sci USA 78:3450–3453

Lacks SA, Greenberg B (1977) Complementary specificity of restriction endonucleases of *Diplococcus pneumoniae* with respect to methylation. J Mol Biol 114:153–168

Lieb M (1983) Specific mismatch correction in bacteriophage lambda crosses by very short patch repair. Mol Gen Genet 191:118–125

Lindahl T (1979) DNA glycosylases endonucleases for apurinic/apyrimidinic sites and base excision repair. Prog Nucleic Acid Res Mol Biol 22:135–192

Lindahl TC (1982) DNA repair enzymes. Ann Rev Biochem 51:61–87

Lindahl T, Nyberg B (1974) Heat-induced deamination of cytosine residues in deoxyribonucleic acid. Biochemistry 13:3405–3410

Marinus MG (1976) Adenine methylation of Okazaki fragments in *Escherichia coli*. Mol Gen Genet 128:853–854

Marinus MG (1980) Influence of *uvr D3*, *uvr E502* and *recL 152* mutations on the phenotypes of *Escherichia coli* K12 *dam* mutants. J Bacteriol 141:223–226

Marinus MG (1982) DNA methylation in *Escherichia coli*. In: Usdin E, Borchardt RT, Cleveling CR (eds) Biochemistry of *S*-adenosylmethionine and related compounds. Macmillan, London, pp 249–253

Marinus MG, Morris NR (1973) Isolation of deoxyribonucleic acid methylase mutants of *Escherichia coli* K12. J Bacteriol 114:1143–1150

Marinus MG, Morris NR (1974) Biological function for 6-methyladenine residues in the DNA of *Escherichia coli* K12. J Mol Biol 85:309–322

Marinus MG, Morris NR (1975) Pleiotropic effects of a DNA adenine methylation mutation (*dam*-3) in *Escherichia coli* K12. Mutat Res 28:15–26

Marinus MG, Konrad B (1976) Hyper-recombination in *dam* mutants of *E. coli* K12. Mol Gen Genet 149:273–277

Mc Graw BR, Marinus MG (1980) Isolation and characterization of dam^+ revertants and suppressor mutations that modify secondary phenotypes of *dam-3* strains of *E. coli* K12. Mol Gen Genet 178:309–315

Meselson M, Pukkila P, Rykowsky M, Peterson J, Radman M, Wagner R, Herman G, Modrich P (1980) Methyl-directed mismatch repair: a mechanism for correcting DNA. J Supramol Struct [Suppl] 4:311

Pukkila PJ, Peterson J, Herman G, Modrich P, Meselson M (1983) Effects of high levels of DNA adenine methylation on methyl-directed mismatch repair in *Escherichia coli*. Genetics 104:571–582

Radman M, Wagner RE, Glickman BW, Meselson M (1980) DNA methylation, mismatch correction and genetic stability. In: Alacević M (ed) Progress in environmental mutagenesis. Elsevier, Amsterdam, pp 121–130

Radman M, Dohet C, Bourguignon MF, Doubleday OP, Lecomte P (1981) High fidelity devices in the reproduction of DNA. In: Seeberg E, Kleppe K (eds) Chromosome damage and repair. Plenum, New York, pp 431–445

Wagner RE, Meselson M (1976) Repair tracts in mismatched DNA heteroduplexes. Proc Natl Acad Sci USA 73:4135–4139

Yuan R (1981) Structure and mechanism of multifunctional restriction endonucleases. Annu Rev Biochem 50:285–315

The *mom* Gene of Bacteriophage Mu

R. KAHMANN

1 Introduction

The idea that gene activity in eukaryotes can be controlled by DNA methylation has gained considerable support over the past years. Examples where
genes are regulated by methylation show an inverse correlation between
the degree of DNA methylation at specific sites and the extent of transcription (for reviews see RAZIN and RIGGS 1980; DOERFLER 1981). A variety
of sequence-specific methylases have been characterized from prokaryotic
organisms. Where these methylases are part of restriction modification systems they serve to protect the host DNA against destruction by the indige-
nous nuclease activity. Where this is not the case, where modification functions exist without a corresponding restricting activity on phage genomes
or as chromosomal genes (see HATTMANN 1981), knowledge about their
biological role is rather poor. The temperate bacteriophage Mu encodes
such a modification function which has been termed Mom (modification
of Mu; TOUSSAINT 1976). This function, though dispensable for phage
growth (TOUSSAINT 1976), is fascinating in two respects. First, the modifica-

Max-Planck-Institut für molekulare Genetik, Ihnestraße 63–73, D-1000 Berlin 33 (West)

tion introduced in DNA is not a methylation, but rather a novel type (HATT-
MAN 1979). Second, the regulation of *mom* gene expression follows a complex
scheme without precedence in prokaryotes: apart from the *mom* gene itself,
another Mu phage-encoded transacting function and the *Escherichia coli*
host deoxyadenosine methylase (Dam) are prerequisites for Mom expression. The levels of Mom-specific modification are further influenced by
growth conditions of the phage (TOUSSAINT 1976). This whole scenario has
been puzzling investigators in the field for a number of years. Only very
recently have approaches in sequencing and cloning begun to unravel the
molecular basis of this system. A most exciting outcome of these experiments
demonstrated a linkage between methylation of specific sequences upstream
of the *mom* gene and its transcription (KAHMANN 1983; HATTMAN et al.
1983; PLASTERK et al. 1983). The purpose of this review is to cover what
is known about the *mom* gene of bacteriophage Mu and to analyze critically
the recent investigations pertaining specifically to the molecular mechanism
by which the *mom* gene is regulated.

2 Mom-Specific Modification

2.1 Biological and Physical Assays

Phage Mu *mom*$^+$ is rather insensitive to several in vivo restriction systems
(TOUSSAINT 1976). This is particularly apparent when plating efficiencies
of phage grown by induction, by infection, or spontaneously released from
lysogens are compared on *E. coli* strains carrying various restriction-modification specificities (hsA, hsB, hsK, hsP). Since the hsP system is the most
commonly used, only these values are given here: Mu *mom*$^+$ grown
by induction or spontaneously released from a lysogen has an
eop $\left(\dfrac{\text{titer on } E.\ coli\ \text{hsP}}{\text{titer on } E.\ coli\ \text{hsp}^-} \right)$ of 0.1–0.7, Mu *mom*$^+$ grown by infection has
an eop of 5×10^{-2}–5×10^{-3}, and Mu *mom*$^-$ has an eop of 10^{-4}–10^{-5},
irrespective of its growth conditions. This dramatic difference in plating
efficiency of Mu *mom*$^+$ and Mu *mom*$^-$ phage is widely used to test *mom*
gene function. Since Mom-specific modification is not restricted to Mu DNA
but acts in trans to modify cellular DNA sequences (TOUSSAINT 1976), cells
expressing the Mom function also modify superinfecting phages such as
λ, induced λ prophages, or plasmids present in the same cell. *Mom*$^+$-modi-
fied λ, like Mu, is less sensitive to in vivo restriction systems (TOUSSAINT
1976). *Mom*$^+$-modified plasmids can be tested by transforming crude plasmid DNA preparations into restricting and nonrestricting strains and comparing transformation efficiencies (HATTMAN et al. 1983). The use of λ or
plasmids allows the assay of Mom-specific modification without introducing
another Mu prophage into the system, e.g., to test constitutive expression
of plasmid-cloned *mom* genes. Mu *mom*$^+$ DNA is at least partially resistant
to the action of a large number of type-II restriction endonucleases in vitro,
while Mu *mom*$^-$ DNA is sensitive. Mom-specific modification in Mu DNA

protects all *Hga*I, *Pvu*II, and *Sal*I sites from cleavage and allows only partial cleavage products to be formed by enzymes *Acc*I, *Alu*I, *Ava*I, *Bal*I, *Bbv*I, *Dde*I, *Hgi*AI, *Hind*II, *Hinf*, *Hph*I, *Mbo*II, *Mnl*I, *Pst*I, and *Taq*I. Cleavage of DNA with enzymes *Asu*I, *Ava*II, *Bam*HI, *Bgl*I, *Bgl*II, *Bst*EII, *Cla*I, *Dpn*I, *Eco*RI, *Hae*II, *Hae*III, *Hha*I, *Hind*III, *Hpa*I, *Hpa*II, *Kpn*I, *Mno*II, *Pvu*I, *Rsa*I, *Sau*3A, and *Sfa*NI is unaffected by Mom-specific modification (R. Kahmann, D. Kamp, in preparation). The large number of *Hga*I cleavage sites present on Mu DNA and most plasmids used as cloning vehicles makes this enzyme extremely useful for detecting even low levels of Mom-specific modification in DNA.

2.2 Nature of the Modified Base

Considering the finding that the *mom*$^+$ phenotype requires both an active *mom*$^+$ gene of Mu and an active *dam*$^+$ gene of the *E. coli* host, it was suspected that the Mom-specific modification might be a methylation of adenine residues at sites not recognized by the *E. coli dam*$^+$ methylase (M. *Eco dam*) alone. The *mom*-gene product could interact with and alter the sequence specificity of the Dam-methylase or increase its activity (TOUSSAINT 1977; KHATOON and BUKHARI 1978). This idea proved to be incorrect when it was shown by paper chromatography that the methyladenine contents of Mu *mom*$^+$ and Mu *mom*$^-$ DNA are similar (HATTMAN 1979). At the same time Hattman noticed that in Mu *mom*$^+$ DNA a significant fraction ($\sim$15%) of the adenine residues are modified to an unusual new form which he termed A_x. A_x is resistant to perchloric acid hydrolysis and to alkaline phosphatase. From its chromatographic and electrophoretic properties it was concluded that A_x contains a free carboxyl group (HATTMAN 1979). Recent studies indicate that A_x is N^6-carboxy-methyladenine, but that this compound might be derived from the more acid labile residue N^6-(1-acetamido)-adenine (D. Swinton, S. Hattman, P.F. Crain, D.L. Smith, C.-S. Cheng, and J.A. McCloskey, in preparation). At present the mechanism that converts adenine to N^6-(1-acetamido)-adenine is unknown. There is no information on which compound might be considered the donor for the acetamido group nor is there knowledge about enzymatic requirements and the number of steps involved.

2.3 Sequence Specificity

Since a large number of restriction endonucleases exhibiting different sequence specificities is affected by Mom-specific modification this was originally taken to indicate a lack of sequence specificity. This assessment, however, did not withstand a more careful analysis. When all known recognition sequences of endonucleases whose cleavage is affected by Mom are compared and aligned (Table 1), a consensus recognition sequence for Mom can be deduced: $5'-\frac{C}{G}A\frac{G}{C}N\,Py\text{-}3'$ (R. Kahmann, D. Kamp, in preparation). In this analysis it was taken into account that in those cases where

Table 1. Mom-mediated protection against cleavage by restriction endonucleases[a]

Degree of Protection of Mu *mom*+ DNA	Restriction enzyme	Number of cleavage sites in Mu DNA	Recognition sequences (1 2 3 4 5)	Remarks
+ + + +	*Hga*I	> 50	G A C G C	
+ + + +	*Sal*I	1	GT C G A C	
+ +	*Acc*I	4	GT C G A C	Subset of GT($_C^A$)($_T^G$)AC
+ +	*Alu*I	> 50	A G C T	
+	*Ava*I	3	C Py C G A G	Subset of CPyCGPuG
+	*Bal*I	3	T G G C C A	
+ + +	*Bbv*I	> 50	G C A G C	Subset of GC($_A^T$)GC
+	*Dde*I	> 30	C T N A G	
?	*Eco*B	?	T G A (N)$_8$ T G C T	In vivo tests
?	*Eco*K	?	G C A C (N)$_6$ G T T	In vivo tests
?	*Eco*PI	?	A G A C Py	In vivo tests
+ +	*Hgi*AI	> 12	G A G C T C	Subset of G($_T^A$)GC($_T^A$)C
			G ($_T^A$) G C A C	
+ +	*Hind*II	17	GT C G A C	Subset of GTPyPuAC
+	*Hin*f	> 50	G A N T C	
+	*Hph*I	> 50	G G T G A	
			T C A C C	
+	*Mbo*II	> 50	G A A G A	
+ +	*Mnl*I	> 50	G A G G	
+	*Pst*I	2	C T G C A G	
+ + + +	*Pvu*II	2	C A G C T G	
+ +	*Taq*I	> 50	T C G A	
+ +	*Xho*II	> 14	A G A T C Py	Subset of PuGATCPy
		Consensus sequence:	$_C^G$ A $_C^G$ N Py	

[a] This table compiles data from ALLET and BUKHARI (1975), TOUSSAINT (1976), KHATOON and BUKHARI (1978), and R. Kahmann and D. Kamp (in preparation)

+, protection of about 25% of all cleavage sites by that particular restriction endonucleases restriction enzymes cleave a degenerate sequence only one subset of this sequence might overlap a Mom site and hence be protected. This can indeed be recognized in the fragmentation pattern of Mu *mom*$^+$ DNA as the specific loss of certain fragments while others appear with the same intensity as in unmodified Mu DNA. The analysis of dinucleotides containing A$_x$ has revealed that only C and G appear to be nearest neighbors of A$_x$ (HATTMAN 1980), in agreement with the proposed recognition site. The small number of cleavage site sequences which extend through the entire consensus sequence makes the assignments of N and Py to positions 4 and 5 of the Mom recognition site rather tentative. The assumption that Mom recognizes a pentamer and not a shorter sequence is based mainly on the notion that all *Hga*I sites (GACGC) in modified DNA are protected, whereas only partial protection of *Mnl*I (GAGG) and *Bbv*I (GCAGC) sites, which are subsets of the consensus sequence up to position four, is observed. Furthermore, assuming a random distribution of bases, the expected value of modified adenine residues amounts to 12.5% which is in close agreement with the measured 15% value for A$_x$ (HATTMAN 1980).

3 Primary Structure of the *mom* Gene

3.1 Genetic Map

Several *mom* mutants of phage Mu were isolated after hydroxylamine mutagenesis and shown to map to the right of a mutation in gene *S* by performing standard crosses with Mu prophage deletion strains (TOUSSAINT 1976). The map location of the *mom* gene has subsequently been refined. Using mini-Mu phages that have deleted large internal parts of the Mu genome the *mom* gene was mapped in the 1700-bp β segment of Mu DNA (TOUSSAINT et al. 1980). Analysis of a series of deletion substitution mutants in the β region (CHOW et al. 1977) permitted genetic separation of the *mom* gene from the only other known gene in the β region, *gin*, which encodes a function responsible for inversion of the G segment (KAMP et al. 1978). Mu 445-8, a mutant which has an intact left half of the β region but has deleted the right portion of β up to position 150 (position 1 defines the right end of Mu) has a *gin*$^+$ *mom*$^-$ phenotype. This places the *mom* gene to the right of, or overlapping, *gin* and makes it the rightmost gene known in Mu (R. Kahmann, D. Kamp, in preparation).

3.2 Physical Map

A cleavage map has been constructed using conventional restriction-endonuclease mapping procedures on cloned DNA segments containing the right end of Mu DNA (Fig. 1). The map covers the region from the right Mu attachment site (attR) to a *Hind*II cleavage site located in *gin*. Noteworthy as landmarks for the chapter on regulation of the *mom* gene are two *Pvu*I and two *Cla*I cleavage sites between positions 1025 and 1060 and a *Bcl*I cleavage site at position 890.

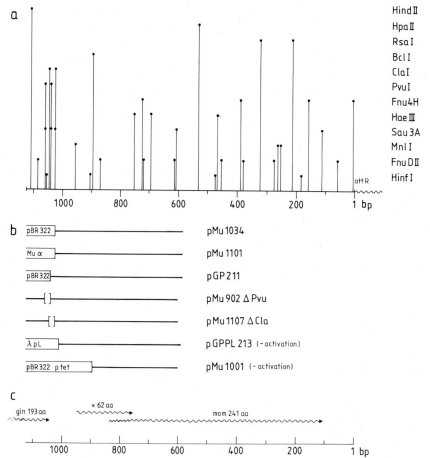

Fig. 1 a–c. Physical map of the *mom* gene and of mutants affecting *mom*-gene expression. **a** Restriction enzyme cleavage map. Lines of the same length correspond to cleavage sites by the endonuclease indicated on the right. *attR* marks the right attachment site of Mu DNA. Numbering in base pairs (*bp*) is from right to left. **b** Physical map of plasmids containing the *mom* gene and alterations introduced in the regulatory region which render Mom expression Dam independent. The maps of pMu1034, pMu1101, pMu902ΔPvu, pMu1107ΔCla, and pMu1001 are taken from KAHMANN (1983) those of pGP211 and pGPPL213 from PLASTERK et al. (1983). *Solid lines* represent Mu DNA. All plasmids contain sequences 600 through 1, which is not indicated here. *Boxes* indicate the origin of sequences that have been fused to Mu DNA. *Brackets* indicate deletions. **c** Transcripts of the *β* region. The length of putative transcripts of the *β* region (deduced from the DNA sequence) is indicated. *Arrows* mark the direction of transcription. *Numbers* refer to the amino acid coding capacity of individual transcripts

3.3 Nucleotide Sequence

The rightmost 1110 bp of Mu DNA containing the *mom* gene have been sequenced (KAHMANN 1983; Fig. 2). This region contains three open reading frames of substantial length in the direction toward the end of the Mu genome (Mu attR). Reading frame 1, corresponding to the *gin* gene, is

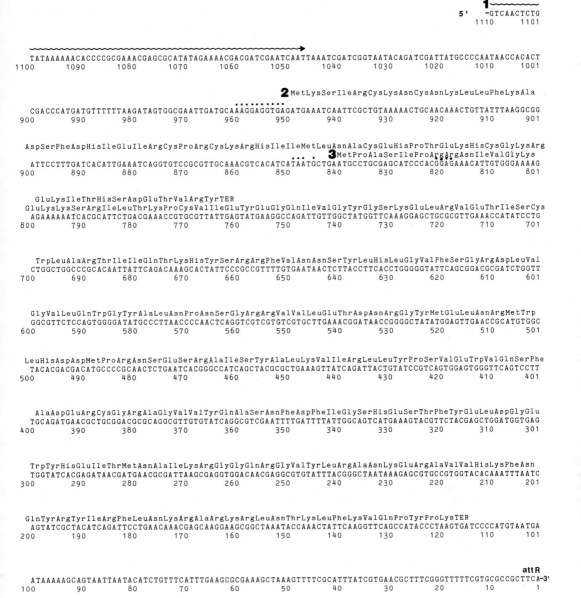

Fig. 2. Nucleotide sequence of the rightmost 1110 base pairs of Mu DNA. Sequence is presented in the 5' to 3' direction. Position *1* marks the first nucleotide of Mu attR (KAHMANN and KAMP 1979). Large open reading frames are indicated. The location of the *gin* gene has been determined previously (R. Kahmann, R. Sandulache and D. Kamp, in preparation). Potential complementarity between the 3' end of 16S RNA (SHINE and DALGARNO 1974) and transcripts of the *β* region is indicated by *dots*

not shown in its entirety. The termination codon for *gin* is located at position 1043 (R. Kahmann, D. Kamp, in preparation). Reading frame 2 (186 nucleotides) starts with an AUG codon at position 949 and reaches an UGA stop codon at position 763. Reading frame 3 could comprise 723 or 693 nucleotides depending on whether AUG at position 840 or GUG at position 810 is used for initiation. An excellent Shine Dalgarno sequence (1974) is found preceding reading frame 2 and a good one near the GUG start of frame 3, only a poor match is found in the region upstream of the AUG start of frame 3 (Fig. 2). The *β* region to the right of the *gin* gene hence could code for two proteins of 62 and 241 amino acid residues with calculated molecular weights of 7.4 and 28.3. The 28.3 value could in fact be lower if the GUG codon at position 810 is used for initiation. The *mom* gene product has not unambiguously been identified on gels, therefore its exact molecular weight remains unknown. The question which of the two reading frames derived from the DNA sequence corresponds to the *mom* gene has been answered in several ways:

1. A Tn5 insertion in the right half of reading frame 3 causes a *mom*⁻ phenotype (D. Kamp, personal communication).
2. A deletion of sequences up to the *Bcl*I cleavage site in pMu1001 (Fig. 1), which removes 60 nucleotides of coding region for the 7.4-K protein and fuses reading frame 3 to the ptet promoter of pBR322, leads to constitutive expression of the Mom function (KAHMANN 1983).
3. An insertion-deletion mutant of Mu, Mu_{445-5} (CHOW et al. 1977), which has deleted part of the *gin* gene and retained instead an IS2 element in the middle of its *β* region (CHOW and BROKER 1978), becomes *mom*⁺ in a *rho*⁻ background (HATTMAN et al. 1983), indicating that Mu_{445-5} contains an intact structural gene for *mom*. Sequence analysis of the Mu_{445-5} mutant, in fact shows, that the IS2 sequences are separated by eight base pairs from the AUG initiation codon of reading frame 3 (R. Kahmann, D. Kamp, unpublished).

Taking these lines of evidence together it must be concluded that the 28.3-K protein (or its shortened version) corresponds to the *mom* gene product. Whether the 7.4-K peptide is made in vivo and what function it might serve remains to be elucidated.

4 Regulation of the *mom* Gene

4.1 The Dam Requirement

When Mu *mom*⁺ phage is propagated in *E. coli* strains which are defective in N-6-deoxyadenosine methylation (*dam*⁻) the resulting phage progeny is modified to an extent about 100 times lower than the same phage propagated in *dam*⁺ strains (TOUSSAINT 1977). Such phage progeny is not only sensitive to in vivo restriction but its DNA can be cleaved to completion with all restriction endonucleases that are usually affected by Mom-specific

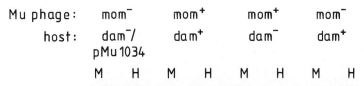

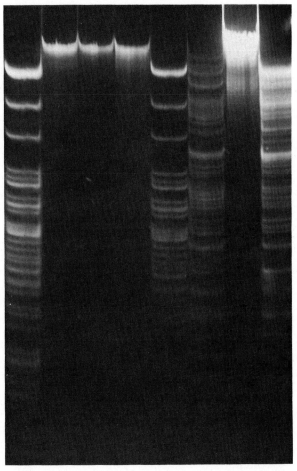

Fig. 3. The presence of Mom- and Dam-specific modification in Mu DNA. The *mom* genotype of the Mu phage is indicated in the *top row*. Phage was propagated by induction in either *dam*[+] or *dam*[−] hosts. The presence of Mom-specific modification was determined by cleavage of the DNA with *Hga*I (*H*). Phage DNA carrying Mom-specific modification is resistant to cleavage with *Hga*I, while unmodified Mu DNA is sensitive (R. Kahmann and D. Kamp, in preparation). *Mbo*I (*M*) cleavage of DNA discriminates between *dam*[+] and *dam*[−] DNA. *Dam*[+] DNA is resistant while *dam*[−] DNA is sensitive to *Mbo*I cleavage. In channels 1 and 2 Mu DNA was isolated from a *dam*[−] host harboring plasmid pMu1034 (see Fig. 1)

modification (Fig. 3; KHATOON and BUKHARI 1978; R. Kahmann and D. Kamp, in preparation). This clearly demonstrates that an active Dam function is a prerequisite for Mom expression.

At the time those experiments were done three reasonable explanations for this finding were discussed:

a) A protein-protein interaction between the *dam* and *mom* gene products might result in a complex that could carry out the Mom-specific modification;
b) *dam*$^+$ DNA could be the substrate for Mom;
c) *dam*$^+$ methylation might regulate the expression of the *mom* gene.

Possibility (a) became highly unlikely when HATTMAN (1979, 1980) showed that Mom neither altered the activity nor changed the specificity of Dam. Recent experiments by KAHMANN (1983) and PLASTERK et al. (1983) demonstrate that the Mom function can be expressed in *dam*$^-$ strains when specific DNA sequences located upstream of the *mom* gene are deleted. This ruled out possibilities (a) and (b), and strongly supported the idea that Dam methylation might regulate the expression of the Mom function. What is known about the details, e.g., sequences and sites involved in this regulatory process, will be discussed in Sect. 3.3. The question on which level the Dam methylase exerts its regulatory role has recently been addressed by HATTMAN (1982). Using a DNA probe containing the *mom* gene he compared transciption of the Mu *mom* gene in *dam*$^+$ versus *dam*$^-$ cells. In a *dam*$^-$ host he observed a 20-fold reduction of *mom* gene transcripts, whereas synthesis of other Mu transcripts showed the same profile as in *dam*$^+$ hosts. This result strongly suggested that the positive regulatory function of Dam acts on the transcriptional level. In this capacity to stimulate transcription of the *mom* gene the Dam methylase can be replaced by the T4 methylase, an enzyme which has the same sequence specificity as Dam but drastically differs from Dam by virtue of its size (S. Hattman, personal communication).

4.2 The Transactivating Function

When DNA fragments containing the *mom* gene are cloned on plasmids they do not usually express the Mom function in the absence of Mu. However, when a Mu *mom*$^-$ prophage is induced in the same cell, the phage DNA is modified (CHACONAS et al. 1981; KAHMANN 1983; PLASTERK et al. 1983). This suggested that the induced prophage provides a function which stimulates the expression of the cloned *mom* genes in trans, or is involved in the modification process itself. The gene encoding this function was identified by testing a series of amber mutations in essential genes for their ability to express Mom. Among these only A*am*, B*am* and C*am* mutants were unable to modify the DNA of a superinfecting λ phage (PLASTERK et al. 1983). Genes *A* and *B* are essential for phage replication, and in their absence the late functions of Mu are not expressed. *C*, on the other hand, is expressed after the onset of replication and is involved in activating the late gene functions of Mu (WESTMAAS et al. 1976). Phages with C*am* mutations replicate normally, indicating that replication per se is not sufficient to activate the *mom* gene. PLASTERK et al. (1983) suggest, therefore, that the *C* gene product itself is the additional Mu function required. Since the formal possibility that the *Cam* mutant used in these experiments carries an additional mutation in the *mom* gene has not been tested, I consider these data suggestive but not yet convincing. Whether the transactivating function interacts with the *mom* gene product in the actual modification process, or rather positively regulates the expression of the *mom* gene has been addressed by fusion of the *mom* gene to foreign promoters (Fig. 1). Under those conditions the *mom* gene can be constitutively expressed (KAHMANN 1983); moreover, when *mom* is transcribed from a foreign promoter the need for the transactivator ceases (PLASTERK et al. 1983). Hence, the role of the transactivating function, like the role of Dam, is regulatory.

4.3 Regulatory Signals

The coding regions for the two known proteins that map in *β, gin* and *mom*, are separated by approximately 200 bp (Fig. 2). This area is extremely rich in secondary structure, characteristic of which are long, direct and inverted repeats clustered around positions 1040 and 940 (see Fig. 4). The respective regions have been termed I and II (KAHMANN 1983).

A striking aspect of region-I sequences is a cluster of three Dam sites in close proximity. At the time I disclosed this structure I felt tempted to speculate that region I might contain the *mom* promoter, and that methylation of the Dam sites and methylation-induced destabilization of potential secondary structure might activate that promoter. What has remained of those speculations? The investigations addressing this question agree on two counts:

1. The *mom* promoter cannot be located within the G segment. Phages having their G segment in either orientation are modified to the same extent (A. Toussaint, personal communication; R. Kahmann and D. Kamp, in preparation) and plasmids carrying the Mu *β* region and the leftmost or the two rightmost thirds of the G segment show identical levels of *mom* expression (KAHMANN 1983).
2. The *mom* promoter is not identical to the *gin* promoter. *Gin* is expressed by the repressed prophage (KAMP et al. 1978), while all attempts to demonstrate this for *mom* have failed (TOUSSAINT 1976).

This leaves three alternatives: the *mom* promoter could lie within the *gin* gene (e.g., upstream of region I), coincide with region I, or be located downstream of region I. Surprisingly, all three options have gained support in recent investigations (HATTMAN et al. 1983; PLASTERK et al. 1983; KAHMANN 1983). All studies pertaining to the subject were carried out with *mom* genes cloned on plasmids. This greatly simplified the introduction of specific changes in the regulatory region and furthermore permitts a direct comparison of results. Plasmids carrying an intact *β* region or sequences 1–1110 (*Hind*II cleavage site, Fig. 1) show a normal pattern of *mom* expression, e.g., require the Dam function and transactivation (KAHMANN 1983; PLASTERK et al. 1983). This places the *mom* promoter to the

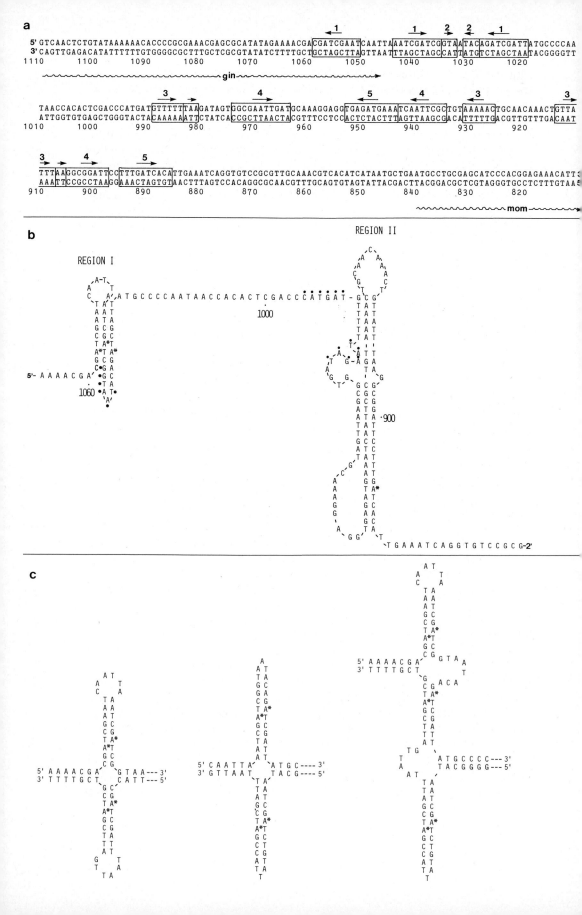

right of the *Hin*dII cleavage site. The region directly preceding the *mom* gene has been subjected to a variety of alterations which can be divided into three classes according to their effects on Mom expression:

1. Small internal deletions in region I (pMu902ΔPvu and pMu1107ΔCla; Fig. 1) which do not affect the surrounding sequences
   By virtue of these deletions either one or two of the three Dam sites in region I have been removed. In these plasmids the expression of Mom is no longer Dam dependent but retains the transactivation requirement (KAHMANN 1983).
2. Deletions which extend into region I, such as those found in pGP211, pMu1034, and pMu1101 (Fig. 1)
   In these plasmids either one or all three Dam sites are removed, and the remaining Mu sequences are fused to sequences of either pBR322 or Mu α origin. In this class of plasmids Mom expression is no longer Dam dependent but requires activation (Figs. 2 and 3; KAHMANN 1983; PLASTERK et al. 1983).
3. Deletions extending beyond the rightmost *Cla*I site in region I, e.g., pMu1001 and pGPPL213 (Fig. 1)
   In these plasmids the Mu sequences have been fused to foreign promoters. Mom expression from these promoters is effective without activation and in the absence of the Dam function. The same fragments cloned next to plasmid sequences without promoter activity fail to express Mom (KAHMANN 1983; PLASTERK et al. 1983).

These studies clearly show that a specific site determines the Dam dependence of Mom expression, and that this site is located in region I. The site is not identical to and can be separated from the site which determines transactivation dependence. The target site for transactivation dependence must lie between the rightmost *Cla*I and the *Bcl*I site, since the *mom* gene in plasmids which have deleted sequences to the left of the *Bcl*I site can no longer be activated (KAHMANN 1983). PLASTERK et al. (1983) present evidence that plasmid pGP213 which contains the same Mu sequences as pGPPL213 but lacks the pL promoter (Fig. 1) has lost the target for transactivation. Unfortunately the exact endpoint of this deletion has not been determined by DNA sequence analysis. If their size measurements are exact, the target for the transactivator must lie between sequences 1026 and 1012, or overlap position 1012 (Fig. 4).

◄ **Fig. 4a–c.** The regulatory region preceding the *mom* gene. **a** Nucleotide sequence of the regulatory region. Regions of dyad symmetry are indicated by *boxes*. *Arrows* indicate that a specific region (referred to by *numbers*) is present as direct (→ →) or inverted (← →) repeat. **b** Hypothetical DNA structures in the regulatory region. To visualize the arrangement of direct and inverted repeats, the DNA is presented as single-strand stem-loop structure. Regions I and II mark areas where secondary structure is most prevalent (KAHMANN 1983). Dam sites are indicated by *asterisks*. Sequences matching the consensus −10 region of known promoter sequences (SIEBENLIST et al. 1980) are indicated by *dots*. **c** Hypothetical secondary structures in region I. Alternative structures which could exist in double-stranded DNA are illustrated. *Asterisks,* as in **b**

How have these data been interpreted, or how should they be, with respect to the location of the *mom* promoter? In the regulatory region preceding the *mom* gene three promoter-like sequences can be found (positions 1036, 995, 983; Fig. 4b) which match known Pribnow boxes (SIEBENLIST et al. 1980), but for all three, homology with the −35 region is rather poor (PLASTERK et al. 1983; KAHMANN 1983). If the transactivating function is a positive effector for turning on transcription of the *mom* promoter by binding to the −35 region, as, for example, the lambda *c*II protein (SHIMATAKE and ROSENBERG 1981), the −35 region need not resemble other promoter sites. Taking these considerations into account PLASTERK et al. (1983) argue that the *mom* promoter is located in region I (position 1036). Since deletion mutants in this region (pMu1034 and pMu1107ΔCla, Fig. 1) retain a promoter which requires transactivation (KAHMANN 1983), the promoter-like sequence at position 1036 cannot coincide with the *mom* promoter. The genuine *mom* promoter must be located to the right of this site, with some probability at position 995 or 983, for instance. A clear distinction between these two putative sites cannot be made with the data available. As I have argued, the target site for the transactivator must lie between sequences 1026 and 1012, or overlap position 1012. Since this region corresponds with the −35 region for the putative promoter at position 995, this site is the most likely candidate for the *mom* promoter. It is apparent that further investigations will be required to firmly establish this proposal.

Recent experiments indicate an even higher complexity of *mom* gene regulation which involves region II (Fig. 4). As pointed out before the *mom* gene can be cloned next to the tetracycline promoter P2 (STÜBER and BUJARD 1981; BROSIUS et al. 1982) of pBR322. Mom expression in such plasmids is constitutive though about 20-fold lower than in an induced lysogen (KAHMANN 1983). Attempts to clone either *Cla*I or *Hin*dII fragments containing the *mom* gene next to plasmid promoters P1 and P2 (STÜBER and BUJARD 1981) respectively were unsuccessful. This failure could not be attributed to the presence of certain sequences on the cloned fragments which might trigger deletions since plasmids containing this region could be isolated when the *mom* gene was inactivated by mutation (R. Kahmann, K. Altmann, unpublished). This suggested that in such plasmids with a wild type *mom* gene high levels of Mom were expressed which caused cell death. This assertion was substantiated by cloning the same fragments next to the λ pL promoter. Such cells die rapidly when the pL promoter is induced (R. Kahmann, K. Altmann, unpublished). These experiments demonstrated furthermore that in plasmids containing the region between the *Cla*I and *Bcl*I site in addition to sequences downstream of the *Bcl*I site expression of the *mom* gene is enhanced compared with plasmids in which only sequences downstream of the *Bcl*I site are present, like in pMu1001 (Fig. 1). It is not yet clear whether this effect has to be attributed to the putative 7.4 K protein which is encoded in this region and which could be seen as an additional positive regulator of *mom* gene expression or whether region II might be involved in stabilizing the *mom* transcript. Since a 9 base pair deletion in the vicinity of the *Bcl*I cleavage site in pMu1034 (Fig. 1) causes a *mom⁻* phenotype (R. Kahmann, K. Altmann, unpublished), complementation assays can now be conducted which should allow to discriminate between a *cis* and *trans* effect of region II on *mom* gene expression.

Are there other transcriptional control sequences located in vicinity of the *mom* gene? Mu-lac fusion phages (CASADABAN and COHEN 1979), which retain the rightmost 116 bp of Mu DNA, apparently do not contain any strong transcription termination signals (Casadaban, personal communication) and the nucleotide sequence between the translational stop codon for *mom* and the right attachment site has no features in common with known termination site sequences (ROSENBERG and COURT 1979). Transcripts of the *mom* gene are therefore not likely to be terminated within Mu sequences, but rather at sites in the adjacent bacterial DNA. HATTMAN (1982) estimates that *mom* mRNA is 1100 nucleotides long. Since this RNA shows a broad migration profile on gels, this size estimate does not contradict the previous statement.

The *gin* gene which is located upstream of *mom* in the β region is expressed in the prophage state (CHOW et al. 1977). One could argue, therefore, that transcription from the *gin* promoter should result in low-level expression of the *mom* gene, particularly since it has now been shown by several investigators that fusion of the *mom* gene to foreign promoters results in *mom* gene expression in the absence of the transactivator (HATTMAN et al. 1983; PLASTERK et al. 1983; R. Kahmann and K. Altmann, unpublished). However, all attempts to detect Mom-specific modification in cells lysogenic for Mu have failed (TOUSSAINT 1976; D. Kamp, unpublished). This could reflect the inadequacy of the assay system to detect low levels of Mom-specific modification; alternatively, it might suggest that a transcription barrier exists between the two genes. Priliminary cloning experiments indicate that such a site is located upstream of the *Bcl*I cleavage site (G. Mertens and R. Kahmann, unpublished). A rigorous study of the transcription pattern of the β region in *dam⁺/dam⁻* DNAs, including the precise mapping of 5′ and 3′ ends of the individual transcripts is required to substantiate this possibility.

4.4 Models

PLASTERK et al. (1983) have suggested a model in which methylation in region I stabilizes the formation of one of several alternative structures (Fig. 4c). In response to the lower degree of methylation shortly after replication when the newly synthesized strand is not immediately methylated (GLICKMAN et al. 1978) the *mom* promoter region might become accessible for the transactivator and the RNA polymerase; it is not when both strands are either fully methylated or unmethylated. Though it is attractive, I feel that the basis for this model – the location of the *mom* promoter in region I – has been shown to be inaccurate. The putative promoter can be deleted; all three Dam sites can be deleted without affecting the transactivator-dependent mode of *mom* gene expression. This suggests that the presence of methylated sequences in region I does not affect *mom* promoter activity directly, but rather in an indirect fashion. In this context, "indirect" means that the degree of methylation in region I might either stabilize or induce a certain DNA conformation (several possible duplex structures are shown in Fig. 4c). This, in itself, could permit or prevent access to the *mom* promoter. Alternatively, a certain DNA conformation might effect termination of the *gin* transcript, retain RNA polymerase in a complex, and thereby prevent initiation at the nearby *mom* promoter. Yet another alternative is that during replication binding of the Dam methylase itself could positively affect access to the *mom* promoter.

If the positive effect of region II on *mom* gene expression can be substantiated, the possibility for yet another regulatory pathway, a regulatory cascade, exists: the Dam function and the transacting Mu function could positively regulate the expression of the 7.4 K protein from a promoter close to region I. The 7.4 K protein in turn could activate the genuine *mom* promoter located within the coding region for the 7.4 K protein.

On the basis of our present knowledge – that is, without data on the exact location of sites involved in the regulatory process, without information about possible interactions of proteins with these sites in response to methylation, and without knowledge about methylation-induced changes in region I – I consider model building premature.

It is worthwhile to note here that a third aspect of *mom* gene regulation, the enhancement of expression after prophage induction, appears to be affected by the amount of methylase present in the cell. In strains which overproduce methylase the levels of Mom-specific modification after infection and induction are similar (S. HATTMAN cited by HATTMAN et al. 1983). These authors speculate that methylation levels in mature phage DNA may be lower than in prophage DNA, and that this might cause the difference in Mom-specific modification usually observed. If half-methylated DNA in region I is a prerequisite for expression, after one round of replication the infecting DNA might have completely unmethylated sites which, as in the *dam⁻* situation, blocks expression of the *mom* gene.

5 Biological Role

The Mom function of Mu has been implicated as allowing the phage to overcome host-controlled restriction/modification systems effectively (TOUSSAINT 1976). While this is well established for Mu with the $G(+)$ host-range specificity (KAMP et al. 1978), not a single host has been found in which modification of the $G(-)$ phage helps improve its plating efficiency (D. Kamp, personal communication). Since the Mom function is dispensable for Mu growth (TOUSSAINT 1976) any involvement of Mom in other processes of Mu development can only be marginal. As pointed out in the preceding section, high levels of Mom-specific modification are lethal to the host. Therefore high-level expression of Mom after prophage induction most certainly shuts down cellular functions. The low level of Mom expressed after phage infection might, in fact, provide a way to retain the possibility for lysogenization of the host and its subsequent survival. S. Hattman (personal communication) has shown that Mom lowers the T_m of Mu DNA by about 4° C. It is tempting to speculate that this might affect the initial insertion of Mu DNA into the host chromosome. One could also envision differences in the specificity of Mu integration depending on whether or not the substrate carries Mom-specific modification. As yet there is no evidence supporting such a scheme (R. Kahmann and D. Kamp, in preparation). The distribution of putative Mom sites in the sequenced parts of Mu and *E. coli* DNA is similar (D. Kamp, unpublished), which makes it unlikely that the Mom function is responsible for preventing Mu from integrating into its own DNA during replication. Apart from reducing endonuclease cleavage in vivo, no other clearcut biological role has been found for the Mom function.

6 Related Functions

A temperate phage, D108 is closely related to Mu in its biological properties (GILL et al. 1981). Electron microscopic analyses of heteroduplexes between Mu and D108 DNA have revealed three regions of nonhomology near the left end, a small divergence within the G segment, and a substitution of about 500 bp in the β region of D108 located close to the attachment site (GILL et al. 1981). Because of this latter substitution, which could conceivably have altered or inactivated the *mom* gene, D108 DNA was probed with restriction endonucleases for the presence of Mom-specific modification. D108 expresses a Mom-like function which shows the same sequence specificity as Mu Mom (R. Kahmann and D. Kamp, in preparation). The D108 Mom function also requires the Dam function of *E. coli* and a transactivator. Plasmids carrying the *mom* gene of D108 can be transactivated by a Mu *mom⁻* phage, which indicates a high degree of similarity in the transactivating proteins, as well as in the target sites which are recognized (R. Kahmann, unpublished). Comparable to the situation described for Mu, the Dam requirement for D108 *mom*-gene expression can be overcome by deleting sequences upstream of the gene (R. Kahmann, unpublished).

7 Summary

The methylation-dependent expression of the *mom* gene of bacteriophage Mu is the first example of a linkage between DNA methylation and gene activity in prokaryotes. Methylation of specific sequences located upstream of the *mom* gene is a prerequisite for transcription of the gene. The sites which have to be methylated, however, coincide neither with the *mom* promoter nor with the sequence recognized by the Mu-specific transactivator. It is not yet understood how certain structural features which are implied by the primary sequence contribute to the Dam-dependent initiation of transcription from the downstream *mom* promoter. At least as obscure remain the biological role of the Mom function during phage development and the mechanism by which the acetamido group is introduced in DNA. Although certain features and a rough alignment of important sites for the regulatory process are suggested by the information described, the relevant critical details of the events remain to be elucidated.

Acknowledgments. I am very grateful to those people who contributed unpublished information used in this review, and to my colleagues at the institute for valuable discussions. I appreciate the help of H. MARKERT in the typing of the manuscript.

References

Allet B, Bukhari AI (1975) Analysis of Mu and λ-Mu hybrid DNAs by specific endonucleases. J Mol Biol 92:529–540

Brosius J, Cate RL, Perlmutter AP (1982) Precise location of two promoters for the β-lactamase gene of pBR322. J Biol Chem 257:9205–9210

Casadaban MJ, Cohen SN (1979) Lactose genes fused to exogenous promoters in one step using a Mu-*lac* bacteriophage: In vivo probe for transcriptional control sequences. Proc Natl Acad Sci USA 76:4530–4533

Chaconas G, de Bruijn FJ, Casadaban MJ, Lupski JR, Kwoh TJ, Harshey RM, Du Bow MS, Bukhari AI (1981) In vitro and in vivo manipulations of bacteriophage Mu DNA: Cloning of Mu ends and construction of min-Mu's carrying selectable markers. Gene 13:37–46

Chow LT, Broker TR (1978) Adjacent insertion sequences IS2 and IS5 in bacteriophage Mu mutants and IS5 in a lambda d*arg* bacteriophage. J Bacteriol 133:1427–1436

Chow LT, Kahmann R, Kamp D (1977) Electron microscopic characterization of DNAs of non-defective deletion mutants of bacteriophage Mu. J Mol Biol 113:591–609

Doerfler W (1981) DNA methylation – a regulatory signal in eukaryotic gene expression. J Gen Virol 57:1–20

Gill GS, Hull RC, Roy Curtiss III (1981) Mutator bacteriophage D108 and its DNA: an electron microscopic characterization. J Virology 37:420–430

Glickman B, van den Elsen P, Radman M (1978) Induced mutagenesis in *dam⁻* mutants of *Escherichia coli*: A role for 6-methyladenine residues in mutation avoidance. Mol Gen Genet 163:307–312

Hattman S (1979) Unusual modification of bacteriophage Mu DNA. J Virology 32:468–475

Hattman S (1980) Specificity of the bacteriophage Mu *mom⁺*-controlled DNA modification. J Virology 34:227–279

Hattman S (1981) DNA methylation. In: Boyer PD (ed) The enzymes, vol 14. Academic, New York, pp 517–548

Hattman S (1982) DNA methyltransferase-dependent transcription of the phage Mu *mom* gene. Proc Natl Acad Sci USA 79:5518–5521

Hattman S, Garadia M, Monaghan C, Bukhari AI (1983) Regulation of the DNA modification function *mom* of bacteriophage Mu. Cold Spring Harbor Symp Quant Biol 47:647–653

Kahmann R (1983) Methylation regulates the expression of a DNA-modification function encoded by bacteriophage Mu. Cold Spring Harbor Symp Quant Biol 47:639–646

Kahmann R, Kamp D (1979) Nucleotide sequences of the attachment sites of Mu DNA. Nature 280:247–250

Kamp D, Kahmann R, Zipser D, Broker TR, Chow LT (1978) Inversion of the G DNA segment of phage Mu controls phage infectivity. Nature 271:577–580

Khatoon H, Bukhari AI (1978) Bacteriophage Mu-induced modification of DNA is dependent upon a host function. J Bacteriol 136:423–428

O'Day K, Schultz D, Ericsen W, Rawluk L, Howe M (1979) Correction and refinement of the genetic map of bacteriophage Mu. Virology 93:320–328

Plasterk RHA, Vrieling H, van de Putte P (1983) Transcription initiation of Mu *mom* depends on methylation of the promoter region and a phage-coded transactivator. Nature 301:344–347

Razin A, Riggs AD (1980) DNA methylation and gene function. Science 210:604–610

Razin A, Friedman J (1981) DNA methylation and its possible biological roles. Prog Nucleic Acid Res Mol Biol 25:33–52

Rosenberg U, Court D (1979) Regulatory sequences involved in the promotion and termination of RNA transcription. Annu Rev Genet 13:319–353

Shimatake H, Rosenberg M (1981) Purified λ regulatory protein *c*II positively activates promoters for lysogenic development. Nature 292:128–132

Shine J, Dalgarno L (1974) The 3′terminal sequence of *Escherichia coli* 16S ribosomal RNA: Complementarity to nonsense triplets and ribosome binding sites. Proc Natl Acad Sci USA 71:1342–1346

Siebenlist U, Simpson R, Gilbert W (1980) *E. coli* RNA polymerase interacts homologously with two different promoters. Cell 20:269–281

Stüber D, Bujard H (1981) Organization of transcriptional signals in plasmids pBR322 and pACYC184. Proc Natl Acad Sci USA 78:167–171

Toussaint A (1976) The DNA modification function of temperate phage Mu-1. Virology 70:17–27

Toussaint A (1977) The modification function of bacteriophage Mu-1 requires both a bacterial and a phage function. J Virol 23:825–826

Toussaint A, Desmet L, Faelen M (1980) Mapping of the modification function of temperature phage Mu-1. MGG 177:351–353

Westmaas GC, van der Maas WL, van de Putte P (1976) Defective prophages of bacteriophage Mu. Molec Gen Genet 145:81–87

Wijffelman C, van de Putte P (1974) Transcription of bacteriophage Mu. Molec Gen Genet 135:327–337

DNA Methylation in Friend Erythroleukemia Cells: The Effects of Chemically Induced Differentiation and of Treatment with Inhibitors of DNA Methylation

J.K. Christman

1 Introduction

1.1 DNA Methylation Before HpaII and HhaI

The basic outlines of our knowledge of DNA methylation in higher eukaryotes were established by 1970. WYATT (1951) had convincingly demonstrated the presence of 5-methylcytosine (5-mC) residues in calf thymus DNA. Chargaff and his colleagues (CHARGAFF et al. 1953; CHARGAFF 1955) had demonstrated that 5-mC residues in DNA were not randomly distributed in different chromatin fractions prepared from calf thymus DNA, and had pointed out that such a finding was incompatible with a new model for DNA structure (WATSON and CRICK 1953) which would predict random substitution of 5-mC for cytosine (C) on the basis of its ability to form equivalent hydrogen-bonded structures with guanine (G) residues. SINSHEIMER (1954) had shown that the only dinucleotide in pancreatic DNase digests of calf thymus DNA with significant 5-mC content was (5-mC, G), which contained more than 30% of the total 5-mC residues in the DNA. This observation and the finding that 5-mC occurs with greatest frequency in sequences where it is either flanked by purines or at the 3′ end of a pyrimidine cluster

Departments of Biochemistry and Pediatrics, Mount Sinai School of Medicine, 1 Gustave Levy Plaza, New York, NY 10029, USA

Current Topics in Microbiology and Immunology, Vol. 108
© Springer-Verlag Berlin · Heidelberg 1984

led to the conclusion that the sequence of the primary methylation site in vertebrate DNA is CpG, and that the frequency of methylation of C residues in this sequence is not influenced by the nucleotide to its 5′ side (DOSKOCIL and SÔRM 1982). KORNBERG et al. (1959) suggested that methylated bases did not enter DNA as deoxynucleotides during polymer synthesis, but rather occurred as a result of specific enzymatic modification of nucleotides already incorporated into DNA. FLEISSNER and BOREK (1962) demonstrated the existence of enzymes capable of transferring methyl groups from S-adenosylmethionine (AdoMet) to nucleic acid polymers (tRNA); a few years later enzymes which transferred methyl groups from AdoMet to DNA were identified in prokaryotes (GOLD and HURWITZ 1963; SRINIVASAN and BOREK 1964). The DNA methyltransferase (MTase) of mammalian cells was shown to be a nuclear enzyme, associated with chromatin (BURDON et al. 1967; SHEID et al. 1968), and evidence was obtained which indicated that DNA is methylated soon after synthesis, although the process can continue for several hours after completion of the S phase of the cell cycle (BURDON and ADAMS 1969; EVANS and EVANS 1970; SNEIDER and POTTER 1969).

As soon as it was shown that DNA could be enzymatically modified and that there was a species specificity to such modification, the following basic hypotheses as to the functions of DNA methylation were proposed: (a) achievement of "individuality" by allowing foreign DNA to be distinguished from endogenous DNA; (b) protection from endogenous nucleases; (c) regulation of differentiation of higher organisms by "cueing in or out of existing capacities for enzyme synthesis"; and (d) induction of oncogenic changes through aberrant changes in the patterns and levels of methylation (see SRINIVASAN and BOREK 1964 for an early summary).

The purification and characterization of enzymes involved in host range restriction of phage replication in prokaryotes not only demonstrated the role of DNA methylation in influencing the interaction of enzymes with DNA and protecting it from nucleolytic cleavage (ARBER and LINN 1969; YUAN and MESELSON 1970; BOYER 1971) but also led to development of the highly sensitive techniques which have allowed the cloning and sequencing of specific genes, incidentally providing a new set of tools for examining patterns of methylation in DNA from other organisms. Later studies indicated that DNA methylation in prokaryotes may also influence mutation rates, fidelity of DNA repair, and recombination frequency (MARINUS and MORRIS 1975; GLICKMAN et al. 1978; COULANDRE et al. 1978; KORBA and HAYS 1982). Although analogous roles have been proposed for DNA methylation in eukaryotic cells, the relative ease with which completely unmethylated foreign DNAs can be introduced into vertebrate cells in a functional form argues against the presence of a restriction/modification system in higher eukaryotes (see however, CATO and BURDON 1979 and BROWN et al. 1978). Since experimental evidence for a role of DNA methylation in recombination or mutation in eukaryotes is also lacking, the search for a role of DNA modification in vertebrate systems has focused on the hypothesis that 5-mC residues in DNA could affect transcriptional activity of genes.

1.2 DNA Methylation as a Regulator of Gene Activity in Vertebrate Cells

To explain how postsynthetic methylation of DNA could modify gene expression, it was proposed that enzymatic deamination of 5-mC residues in DNA might act to cause transitional mutations in regulatory regions which would in turn alter gene activity (SCARANO et al. 1967), that gene inactivation might be mediated through a restriction/methylation system which could inactivate unmethylated genes (SAGER and KITCHIN 1975), that methylation of C residues in sequence in specific areas of the genome might act as a "clock," turning genes on at appropriate times during embryonic development (HOLLIDAY and PUGH 1975), or that methylation of specific "inactivation sites" might activate X chromosomes (RIGGS 1975). Although subsequent research indicated that methylation does not activate genes but instead may be a factor in blocking their transcription, these authors were among the first to point out that postsynthetic methylation of DNA could lead to heritable changes in the genome. Our current concepts of how methylation patterns are inherited still rely on their assumption that eukaryotic DNA MTases have properties similar to those of prokaryotic type II restriction MTases, i.e., that methylation occurs at symmetrical sites, $_{-GpC-}^{-CpG-}$; that such sites are usually methylated on both strands, $_{-Gp5-mC-}^{-5-mCpG-}$; and that methylation patterns would be retained because hemimethylated sites $_{---GpC-}^{-5-mCpG-}$ which arose as a consequence of semiconservative replication of DNA would be methylated by MTases with a high degree of preference for substrate sites methylated on only one strand.

There is now an accumulation of experimental data indicating that the process of DNA methylation in vertebrates meets many of the criteria which would be required for an epigenetic determinant of phenotypic characteristics of differentiated cells:

1. Methylation of C residues in DNA alters the physicochemical properties of DNA and affects DNA:protein interactions. Substitution of 5-mC for C in both synthetic oligonucleotides (GILL et al. 1974) and native DNAs (EHRLICH et al. 1975) increases their T_m, indicating that 5-mC:G base pairs are more stable than C:G base pairs. Substitution of 5-mC for C in the synthetic oligonucleotide poly(dG-dC):poly(dC-dG) also allows transition from a right-handed (B) to a left-handed (Z) helix to occur at physiological salt concentrations (BEHE and FELSENFELD 1981). Finally, in addition to reducing the affinity of restriction endonucleases for their recognition/cleavage sites, 5-mC residues have been shown to affect the binding of repressor proteins to regulatory sequences in the *lac*, *ara*, and *lex* genes of *Escherichia coli* (FISHER and CARUTHERS 1979; A. Horwitz and G. Wilcox, R. Brent, cited in KORBA and HAYS 1982).

2. Methylation of C residues in CpG sequences is maintained during DNA replication in vertebrate cells. Studies of DNA sequences methylated in vitro and integrated into mammalian chromosomes after transfection indicate that while 30%–40% of methyl groups in C5-mCGG sites can be lost in a random fashion during prolonged passage of cells (STEIN et al. 1982a), methyl groups in the sequences C5-mCGG and G5-mCGC are generally maintained (POLLACK et al. 1980; WIGLER et al. 1981). Hemimethylated C5-mCGG and G5-mCGC sites in DNAs introduced into vertebrate cells became fully methylated and persisted in a methylated state, while methylation at hemimethylated AG5-mCT, GG5-mC5-mC, 5-MC5-mC(A/T)GG, and 5-mCCGG sites was rapidly lost (STEIN et al. 1982a; HARLAND 1982), demonstrating that maintenance methyltransferases in vertebrate cells are site (CpG)-specific.

3. Methylation of C residues in CCGG sequences can influence gene expression in vertebrate cells. In vitro methylation of herpes thymidine kinase (tk) genes with *Hpa*II MTase decreased the efficiency with which these genes transformed tk mouse cells (POLLACK et al. 1980; WIGLER et al. 1981); the few transformed clones arising after transfection with methylated DNA contained partially unmethylated integrated tk sequences. Under conditions where no selection was made for expression, transfected adenosine phosphoribosyltransferase (aprt) genes were found to integrate with the same freqency regardless of whether they were methylated or not. However, aprt transcripts were found only in cells with integrated unmethylated aprt sequences (STEIN et al. 1982b). In vitro methylation of *Hpa*II sites suppressed expression of adenovirus and SV40 genes introduced into *Xenopus laevis* oocytes by microinjection (VARDIMON et al. 1982; FRADIN et al. 1982), although methylation of A residues in *Eco*RI sites of SV40 DNA (WAECHTER and BASERGA 1982) or general methylation of C residues in SV40 DNA with purified rat liver DNA MTase (SIMON et al. 1982) did not affect expression of viral genes when the DNA was microinjected into nuclei of mammalian cells. These exceptions to the generally observed correlation between methylation of DNA and suppressed gene expression may reflect the fact that methyl groups were transferred to bases which are rarely methylated in vertebrate cells, A residues and C residues in non-CpG sites. However, since *Hpa*II methyltransferase can methylate at most 6% of CpG residues in DNA, it is unlikely that the effects of "natural" patterns of methylation were tested in any of the experiments cited above. The most direct evidence that DNA methylation patterns established in vivo affect gene expression is the demonstration that M-MuLV sequences integrated into the genome of germ line cells in an extensively methylated form are not infectious as isolated but become so when their C residues are unmethylated (HARBERS et al. 1981). Similarly, lymphoid lines persistently infected with herpesvirus contain unmethylated viral genomes; when virus production ceases remaining viral genomes are heavily methylated; when virus production is again induced only unmethylated viral genomes are detected (YOUSSOUFIAN et al. 1982).

4. Active transcription of cellular genes can often be correlated with the presence of unmethylated CpG sites. Examination of a variety of genes by restriction endonuclease cleavage has indicated (a) that there are tissue-specific variations in extent of methylation of CpG-containing restriction endonuclease recognition/cleavage sites; (b) that at least some unmethylated CpG sites present in actively transcribed genes are methylated in DNA from tissues in which the gene is quiescent; (c) that even actively transcribed genes contain many completely methylated CpG sites; (d) that quiescent genes may contain unmethylated CpG sites; and (e) that the pattern of methylation of CpG sites accessible to examination with restriction endonucleases[1] may not change at all when genes are activated or shut off. These apparently contradictory conclusions can be drawn from the early restriction endonuclease studies of WAALWIJK and FLAVELL (1978), McGHEE and GINDER (1979), MANDEL and CHAMBON (1979), and BIRD et al. (1979), and each has been supported by a mass of data derived subsequently in other systems. (For recent reviews see RAZIN and RIGGS 1980; EHRLICH and WANG 1981; DOERFLER 1981.)

5. Methylation levels and patterns vary from tissue to tissue. The studies cited above showed that patterns of methylation of restriction sites in specific genes could vary in a tissue-specific manner. Although early reports of characteristic variations in 5-mC content of DNAs from different tissues (VANYUSHIN et al. 1970, 1973; KAPPLER 1971) conflict with later studies that found tissue variations in 5-mC content too small to be of significance (POLLACK et al. 1978; SINGER et al. 1979), it is clear that differences in 5-mC content of tissues can be convincingly demonstrated by appropriately sensitive chromatographic methods (EHRLICH et al. 1982). However, a comparison of reports on methylation of DNA from bovine sperm indicate that conclusions as to the extent of methylation in a particular tissue are strongly dependent on whether methylation is examined in total DNA, a specific subfraction of DNA, or a specific

---

1 Not all 5-mC residues in DNA of vertebrate cells are found in CpG sequences (SNEIDER 1980), and only a small fraction of CpG sequences can be examined using a given restriction endonuclease with a tetranucleotide ($1/4^2$ or approx. 6%) or a hexanucleotide recognition/ cleavage sequence ($1/4^4$, $>0.4\%$). If all restriction cleavage sites prove to exhibit a strict requirement for palindromic sequences, it can be calculated that a maximum of 25% of CpG sites will be accessible for study

gene, and whether the method employed detects all 5-mC residues or only 5-mC residues in CpG dinucleotides in restriction endonuclease cleavage sites. The 5-mC content of bovine sperm DNA has been found to be up to 50% lower than that of DNA from somatic tissues (VANYUSHIN et al. 1970; STURM and TAYLOR 1981), and it is more sensitive than somatic DNA to cleavage by HpaII, HhaI, and AvaI, indicating that fewer of its CpG sites are methylated (KAPUT and SNEIDER 1979; STURM and TAYLOR 1981). Nevertheless, globin, ovalbumin, ovomucoid, and conalbumin genes are more highly methylated on HpaII and HhaI sites in sperm than in somatic tissues (MANDEL and CHAMBON 1979; WAALWIJK and FLAVELL 1978). Satellite I DNA from bovine sperm has fewer than 1% of its C residues methylated [approximately 10% of the 5-mC content of bovine thymus satellite I DNA (STURM and TAYLOR 1981)], yet is less sensitive to digestion with AvaI (PAGES and ROIZES 1982).

6. Agents which alter cellular phenotypes can affect levels of DNA methylation and agents that act as inhibitors of DNA methylation can alter cellular phenotypes or cause activation of specific genes. Our initial studies with Friend erythroleukemia (FL) cells (CHRISTMAN et al. 1977) indicated that the DNA of these cells becomes hypomethylated as they differentiate and that L-ethionine, which can inhibit in vivo methylation of DNA, acts to trigger differentiation in these cells as well as in a human promyelocytic leukemia cell line (MENDELSOHN et al. 1980). LAPEYRE and BECKER (1979) and LAPEYRE et al. (1981) showed that the level of DNA methylation in hepatocellular carcinomas and premalignant nodules induced in the livers of animals treated with acetylaminofluorene (AAF), diethylnitrosamine, and other carcinogens was markedly lower than in adjacent histologically normal tissues, and AAF bound to DNA decreases its ability to act as a substrate for methylation in vitro (PFOHL-LESZKOWICZ et al. 1981). Persistent hypomethylation of DNA was also noted to occur in Raji cells treated with N-methyl-N-nitrosourea (BOEHM and DRAHOVSKY 1981a).

5-Azacytidine (5-azaCR), which acts in vivo as a specific inhibitor of methylation of C residues in DNA and RNA (LU and RANDERATH 1980; FRIEDMAN 1979) triggers differentiation of $C_3H10T^1/_2$ cells (CONSTANTINIDES et al. 1977; see this volume, Chap. by TAYLOR et al.), FL cells (CREUSOT et al. 1981, 1982; Sect. 3), and HL-60 cells (MENDELSON et al. 1981; CHRISTMAN et al. 1982). Treatment of a wide variety of eukaryotic cells with 5-azaCR has also been shown to activate expression of specific cellular and viral genomes (reviewed in this volume, Chap. by TAYLOR et al.).

It is clear from this summary that there is a strong basis for the hypothesis that DNA methylation may regulate changes in gene expression during vertebrate development. It is just as clear that there is not a simple relationship between the presence or absence of methyl groups in or around a particular gene region and the rate at which the gene is transcribed. As we have speculated previously (CHRISTMAN et al. 1977), it is possible that only a few specific methylation sites need to become "methyl-free" in order to interfere with binding of a repressor protein or to enhance binding of RNA polymerase. However, these "critical sites" for methylation have not yet been completely characterized. Although the possibility that such sites are in promoter regions is suggested by a coordinated appearance of DNAseI-"hypersensitive" sites 5' to genes, a phenomenon associated with increased transcriptional activity, and loss of methyl groups from the hypersensitive region (MCGHEE et al. 1981; GROUDINE et al. 1981), there are also indications that hypersensitive sites and increased transcription can be induced under conditions where it is unlikely that changes in methylation occur (GROUDINE and WEINTRAUB 1982). In addition, there is no evidence from studies of vertebrate systems for the existence of specific DNA:protein interactions affected by the presence of 5-mC residues in DNA. Most strikingly, there is little indication as to how the process of DNA methylation could be regulated during development with the presumed specificity required for establishing different methylation patterns either through de novo methylation of previously unmethylated sites or through directed loss of methyl groups during DNA replication. In an attempt to approach these questions, we have concentrated our studies on the process of DNA methylation in cultured vertebrate cells, which can be triggered to undergo a major shift in phenotype under controlled conditions. This chapter contains a summary of our findings and compares them with findings in other vertebrate systems.

2 Characteristics of DNA Methylation in FL Cells

2.1 Distribution of 5-Methylcytosine Residues in FL Cell DNA

We have examined the extent of methylation of C residues in total FL cell DNA and in FL cell DNA fractionated by a variety of methods (Table 1). FL cell DNA does not contain detectable 6-methyladenine (6-mA) (CHRISTMAN 1982) but approximately 3.7% of its C residues are methylated. The 5-mC content of FL cell satellite DNA is markedly higher than this average value (13.9%), as is the 5-mC content of highly reiterated DNA sequences (5.9%) isolated on the basis of their ability to anneal at a $C_o t < 0.3$. However, C residues in the rapidly annealing fraction of DNA isolated from the main band are no more extensively methylated than C residues in total main band DNA (2.7%). Finally, fewer than 2% of C residues in DNA from chromatin regions which are enriched in sequences undergoing active transcription [operationally defined as DNA from chromatin which can be solubilized by limited digestion with DNase-II and which remains soluble in Mg^{2+} (GOTTESFELD and BUTLER 1977)] are methylated. These results support the hypothesis that regions of the DNA which are not transcribed are more highly methylated than those which are, although it not clear to what extent these results are influenced by variable distribution of CpG dinucleotides among the different fractions.

We have found that approximately 40% of the 5-mC residues in FL cell DNA are flanked by purines and that the rest are distributed in pyrimidine isostichs with a frequency which approaches but is not identical to that predicted if all 5-mC residues are found in the sequence $Pu(Py_n)$5-mCPu (CHRISTMAN 1979). This result is consistent with CpG as the major site of methylation, and indeed, ~70% of the penultimate C residues in MspI (CCGG) sites and 48% of C residues in TaqI (TCGA) sites in FL cell DNA are methylated. Thus the level and distribution of 5-mC in FL cell DNA is basically similar to that found for other mouse DNAs: 3.6% of C residues in P815 mastocytoma cell DNA are methylated (BOEHM and DRAHOVSKY 1981 b), as are 3.2% of C residues in L929 cell DNA (TURNBULL and ADAMS 1976). 5-mC levels for adult mouse DNA range from 2.74 and 3.34 for liver and spleen respectively (determined by HPLC analysis of bases, SINGER et al. 1979) to 3.9 for liver (determined by paper chromatography, KAPPLER 1971). C residues in satellite DNAs from a variety of mouse

Table 1. Distribution of 5-mC residues in FL cell DNA. (Data compiled from results presented in CHRISTMAN (1979), CHRISTMAN et al. (1980), CHRISTMAN (1982), and B. Schoenbrun and J.K. Christman, unpublished work)

Fraction	% total DNA	% C as 5-mC
Total DNA	100	3.7 ± 0.16
Main band[a]	83	2.7 ± 0.10
Satellite[a]	17	13.9 ± 0.13
Highly repetitive ($C_ot\leq0.3$)[b]	20	5.9 ± 0.19
Moderately repetitive ($C_ot>0.3$)[b] and unique	80	2.6 ± 0.12
Active chromatin[c]	12	2.0 ± 0.15
Highly repetitive active chromatin	2	2.1 ± 0.21
Inactive chromatin[c]	88	3.9 ± 0.14
CpG dinucleotides in sites cleaved by:[d]		
MspI (C/CGG)		70.8 ± 0.94
TaqI (T/CGA)		47.6 ± 0.62

Except for determination of methylation of C residues in restriction endonuclease sites, DNAs were labeled with ^3H-uridine in vivo for 18 h prior to isolation, acid hydrolysis, and base analysis by HPLC on Partasil SCX as described in CHRISTMAN et al. (1980).
[a] Satellite and main band DNA were separated on $AgNO_3CsSO_4$ gradients (LIEBERMAN 1973)
[b] Sheared DNA was annealed to the indicated C_ot and separated into single- and double-stranded fractions by chromatography on hydroxyapatite. Double-stranded DNA was digested with S1 nuclease prior to base analysis
[c] Preparation and digestion of chromatin and separation into Mg^{2+}-soluble and -insoluble fractions as described in GOTTESFELD and BUTLER (1977)
[d] DNA was labeled with ^{32}P at the 5' end after digestion with MspI or TaqI prior to enzymatic hydrolysis using a modification of the procedure of CEDAR et al. (1979). Nucleotides were analyzed by reversed-phase HPLC (CHRISTMAN 1982)

cell lines have been shown to be approximately three times more methylated than those in main band DNA (SALOMON et al. 1969; HARBERS et al. 1974; GANNT et al. 1973), and C residues in highly reiterated DNA to be two- to three times more methylated than unfractionated DNA (BOEHM and DRAHOVSKY 1981b).

Differences in methylation of FL cell DNA and normal mouse DNA which are not reflected as differences in 5-mC content can be detected by restriction analysis. Although FL cell DNA is more readily cleaved by HpaII than is normal mouse spleen DNA (SMITH et al. 1982), the penultimate C residues in CCGG sites are methylated to an equivalent extent (65%–70%) in both mouse and FL cell DNA (CHRISTMAN 1982). Approximately 70% of C residues in TCGA sites of liver DNA are also methylated (J.K. Christman, unpublished observation), yet in FL cell DNA fewer than 50% of C residues in TCGA sites are methylated (Table 1). It cannot be concluded that these differences are unique to FL cells or even to cultured cells until the extent of methylation at CCGG and TCGA sites in other mouse tissues and cells has been examined, but it is clear that the extent of methylation of C residues in CCGG sites is not necessarily representative of the extent of methylation of all CpG sites.

2.2 Characteristics of FL Cell DNA Methyltransferase

The DNA MTase of FL cells is a nuclear enzyme tightly bound to chromatin. FL cell cytoplasm does not contain detectable DNA MTase activity and little or no active enzyme is released from intact nuclei incubated at 37° in low-salt buffer (CREUSOT and CHRISTMAN 1981). It is quantitatively extracted from nuclei or chromatin fractions with 0.3 M NaCl, and in low-salt buffer more than 60% of active enzyme can be released by incubating intact nuclei with micrococcal nuclease under conditions where less than 10% of nuclear DNA is digested. We have reported evidence which indicates that this release occurs because the enzyme is primarily associated with internucleosomal or "linker" DNA and is freed when linker DNA is degraded with micrococcal nuclease. DNaseII, which cuts linker DNA less efficiently than micrococcal nuclease, releases approximately 20% of the active DNA MTase in intact nuclei under conditions where 10% of nuclear DNA is digested, leaving the bulk of enzyme associated with linker DNA in Mg^{2+}-precipitable oligonucleosomes (less than 4% of the total DNA MTase activity is found with Mg^{2+}-soluble "active" nucleosomes). Together with the observation that little enzyme is released from chromatin digested under conditions reported preferentially to release (or degrade) actively transcribed chromatin (DIMITRIADIS and TATA 1980; WEINTRAUB and GROUDINE 1976; GAREL and AXEL 1976), this indicates that the bulk of active DNA MTase is associated with linker DNA in condensed regions of chromatin (CREUSOT and CHRISTMAN 1981). Since chromatin containing newly replicated DNA is highly accessible to nucleases (SEALE 1978), the results also suggest, as did those of ADAMS (1974), that DNA MTase is not associated with DNA immediately after its synthesis.

Our studies of DNA MTase have employed a 0.3 M NaCl extract from lysed FL cell nuclei (for details see CREUSOT et al. 1982) which contains a variety of nonhistone nuclear proteins but is free of DNA (by fluorometric analysis by the method of BRUNK et al. 1979) and histone (F. Creusot, unpublished work). If substrate DNAs are prepared from FL cells or other mammalian cells (CHRISTMAN et al. 1982), the enzyme is most efficient in transferring methyl groups to hypomethylated double-stranded (ds) DNA obtained from cells treated with L-ethionine or 5-azaCR (50–150 pmol CH_3/100 µg; Tables 2 and 3). The distribution of methyl groups transferred to this DNA in vitro is identical to the distribution of methyl groups transferred to it in vivo and thus approaches that predicted if most methylation occurs at $Pu(Py)_nCpG$ sites (CHRISTMAN 1979).

Although some purified mammalian DNA MTases display a preference for single-stranded (ss) DNA substrates regardless of whether they are of eukaryotic or prokaryotic origin (ROY and WEISSBACH 1975; SIMON et al. 1978), the ability of FL cell DNA MTase to methylate homologous ds DNAs in vitro does not indicate that these DNAs must contain ss regions. Pretreatment of hypomethylated ds DNAs with S1 nuclease has no effect on their ability to accept methyl groups in vitro and fewer than 1% of the methyl groups transferred to ds DNAs in vitro can be released by

Table 2. In vitro methylation of DNA from FL cells grown in the presence of inducers and/or inhibitors of differentiation. (After CHRISTMAN et al. 1980)

Treatment of cells prior to DNA isolation	Strain 745A		Clone DR-10	
	pmol CH_3/ 100 µg DNA	% B(+)	pmol CH_3/ 100 µg DNA	% B(+)
None	10.8	1–2	12.3	>1
Me_2SO (240 mM)	41	78	12.8	1–2
Butyrate (2 mM)	36.2	82	ND	
L-Ethionine (4 mM)	54.6	40	36	80
HMBA (4 mM)	30.7	95	40	55
BrdUrd (10^{-6} M)	10	1–2	ND	
BrdUrd + Me_2SO	11.7	1–3	ND	
TPA (1.6×10^{-7} M)	10.5	0–1	ND	
TPA + Me_2SO	15.8	7	ND	

HMBA, hexamethylene-bis-acetamide; BrdUrd, 5-bromo-2′-deoxyuridine; TPA, 12-O-tetradecanoyl phorbol-13-acetate
Assays for methyl transfer are described in CHRISTMAN et al. (1980). All values represent the average incorporation for three to five assays on a minimum of three separate DNA preparations from cells grown in the presence of the indicated inducing agents for 3–4 days. Variations did not exceed ±5 pmol. Values for methylation of DNA from cells grown in the presence of TPA and BrdUrd are from a typical experiment and each represents an average of triplicate determinations on a single DNA preparation. BrdUrd was added to the culture medium 24 h before addition of Me_2SO. TPA and Me_2SO were added simultaneously. Benzidine-positive (B+) cells were scored as described by ORKIN et al. (1979) in cultures exposed to inducing agents alone at 5 days and in cultures exposed to inhibitors or inhibitors in combination with inducers at 4 days (Me_2SO-treated cells had 55% B+ cells in 4 days)

S1 nuclease treatment (CHRISTMAN 1979). Furthermore, ss DNAs, whether from untreated cells or from cells treated with L-ethionine or 5-azaCR, are poor substrates, accepting only 2–5 pmol CH_3/100 µg. These properties are consistent with those of an enzyme that preferentially transfers methyl groups to C residues in hemimethylated sites, and may explain why prokaryotic ds DNAs rich in completely unmethylated CpG sites are not good substrates (*E. coli* and pBR322[1] DNA accept 5–10 pmol CH_3/100 µg). It should be noted, however, that ss *E. coli* DNA is readily methylated by FL cell DNA MTase (60–80 pmol CH_3/100 µg DNA). This indicates that the enzyme can initiate methylation at completely unmethylated CpG sites, but does so most efficiently with an ss DNA substrate which contains no methylated CpG sites. The substrate specificity of FL cell DNA MTase resembles that of Krebs II ascites (TURNBULL and ADAMS 1976; GRUENBAUM et al. 1982) or partially purified Novikoff hepatoma cell (SNEIDER et al. 1975, 1979) DNA MTase.

---

1 Transfer of 10 pmol methyl groups to 100 µg pBR322 DNA is the equivalent of methylating one C residue per 43000 base pairs (ten genomes). There are 314 CpGs per pBR322 genome (SUTCLIFFE 1979)

Table 3. Effect of 5-azaCR and 5-azaCdR treatment of FL cells on methyl acceptance of DNA and the level of DNA methyltransferase isolated from the cells. (Data derived partly from CREUSOT et al. 1982)

Treatment	Concentration (μM)	Time (h)	In vitro methyl acceptance (pmol/100 µg)	DNA methyltransferase content (U/10^8) cells
5-azaCR	1	4	12	21.9
	1	10	30	3.2
	1	20	56	1.2
	10	4	23	2.5
	10	10	60	ND
	10	20	110	ND
5-azaCdR	0.4	4	16	17.4
	0.4	10	48	2.7
	0.4	20	104	>0.5
None			12	29

In vitro methyl transfer to DNA isolated from cells treated with 5-azaCR or 5-azaCdR was determined as described in CREUSOT et al. (1982) using 0.3 M NaCl extract from nuclei of untreated FL cells as enzyme source. DNA methyltransferase activity in nuclear extracts from 5-azaCR- or 5-azaCdR-treated cells was determined using a standard preparation of DNA from FL cells grown in presence of 4 mM L-ethionine for 3 days. One unit of enzyme transfers 1 pmol methyl groups per 15 min to this DNA under assay conditions described in CREUSOT et al. (1982)

2.3 DNA Methylation During Differentiation of FL Cells

One of the earliest indications that hypomethylation of DNA might play a role in allowing activation of gene expression was the observation that L-ethionine acts as an inducer of differentiation in FL cells under culture conditions (4 mM L-ethionine, 0.1 mM methionine) where it has minimal effects on cell growth and metabolism (CHRISTMAN et al. 1977). FL cells exposed to 4 mM L-ethionine displayed increased levels of globin mRNA within 48 h, and within 72 h up to 15% of the cells in the culture had accumulated sufficient heme or hemoglobin to stain positively with benzidine.[1] DNA, tRNA (CHRISTMAN et al. 1977), and chromatin proteins (COPP 1981, unpublished work) isolated from the treated cells were all "hypomethylated" in the sense that they served as better methyl acceptors than comparable preparations from untreated FL cells when FL cell enzymes were

---

1 The changes with occur during FL cell differentiation include accumulation of globin mRNA, α, and β-globin chain synthesis, increase in heme synthesis and in a variety of proteins characteristic of or specific to erythrocytes (catalase, enzymes involved in heme synthesis spectrin, erythrocyte-specific membrane antigens). Cellular morphology changes, chromatin becomes condensed, and purine metabolism is altered (reviewed in MARKS and RIFKIND 1978). The most convenient indication that this extensive program of phenotypic changes has occurred in an individual cell is the accumulation of heme, which can be detected by benzidine staining. The terms "benzidine-positive cell" (B+) and "differentiated FL cell" are commonly used interchangeably

used to catalyze methyl transfer from *S*-AdoMet in vitro. This finding is not surprising, since L-ethionine in the form of *S*-adenosylethionine is a competitive inhibitor of a variety of methyl transfer reactions in which *S*-AdoMet acts as methyl donor. However, it was important evidence that L-ethionine acted to induce FL cell differentiation in the same concentration range in which it acted as an inhibitor of transmethylation. The evidence indicating that L-ethionine acted to induce differentiation by some mechanism related to its ability to inhibit methylation and that the critical reaction was methylation of DNA rather than RNA or protein can be summarized as follows:

1. DNA from FL cells grown in the presence of inducing concentrations of Me$_2$SO, butyrate, hexamethylene-bis-acetamide, and pentamethylene-bis-acetamide is hypomethylated by the criterion of increased ability to accept methyl groups in vitro (Table 2).

2. tRNA (CHRISTMAN et al. 1977 and unpublished work) and chromatin proteins (COPP 1981) from FL cells treated with inducers other than L-ethionine do not differ from tRNA and chromatin proteins of untreated FL cells in their ability to serve as substrates for in vitro methylation by homologous enzymes.

3. Hypomethylation of DNA can be detected as early as 24 h after exposure of FL cells to inducing agents (CHRISTMAN et al. 1977), a time which coincides with the onset of commitment of significant numbers of FL cells to differentiation (MARKS and RIFKIND 1978). Both the fraction of cells committed to differentiation and the degree of hypomethylation of DNA increase with time of exposure to inducing agents.

4. FL cell DNA does not become hypomethylated when the cells are exposed to inducing agents under conditions where differentiation does not or cannot occur. DR-10 cells, subcloned from FL cells (745A), are not induced to differentiate by Me$_2$SO, even though they take up the compound to the same extent and are as sensitive to its toxic effects as are FL cells that can be induced to differentiate (OHTA et al. 1976). DNA from DR-10 cells maintained in the presence of Me$_2$SO does not differ in its ability to accept methyl groups in vitro from DNA of DR-10 cells grown in the absence of Me$_2$SO or from DNA of untreated FL cells of other clones. In contrast, DNA from DR-10 cells grown for 4 days in the presence of agents that can induce their differentiation accepts 35–45 pmol CH$_3$/100 µg DNA, a value quite comparable to those obtained for methylation of DNA from differentiating cells of the parental strain (Table 2).

Two inhibitors of FL cell differentiation, 5-bromo-2-deoxyuridine (BrdUrd) (PREISLER et al. 1973) and 12-*O*-tetradecanoyl phorbol-13-acetate (TPA) (ROVERA et al. 1977; YAMASAKI et al. 1977) do not detectably affect methylation of FL cell DNA in vivo or alter its ability to be methylated in vitro (CHRISTMAN et al. 1980). However, when FL cells are exposed to an inducer of differentiation (Me$_2$SO) in the presence of either TPA or BrdUrd, both differentiation and the increase in in vitro methyl acceptance of DNA normally observed in inducer-treated cells are suppressed (Table 2).

Quantitation of transfer of radiolabeled methyl groups from *S*-AdoMet to DNA in vitro by a homologous DNA MTase is a highly sensitive method for detection of "methylatable" sites, transfer of 1 pmol CH$_3$/100 µg DNA being the equivalent of addition of one methyl group per $\sim$300 000 bases. However, the validity of interpreting an increase in number of methyl groups transferred to DNA in vitro as indicative of an overall loss of methyl groups, i.e., the presence of methylation sites which remained unmethylated in vivo, depends to a large extent on the specificity of the MTase. As described in Sect. 2.2, the FL cell DNA MTase we employed appears to have retained the specificity displayed by the enzyme during methylation of DNA in vivo, preferring homologous hemimethylated ds DNA as substrate and transferring methyl groups to C residues in a nonrandom fashion. It can be calculated from the data in Table 2 that DNA from FL cells undergoing terminal differentiation has $\sim 300\,000$ more methylatable C residues per haploid genome than DNA from untreated cells. If every methyl group accepted in vitro represents transfer of a methyl group to a C residue that would have been methylated in vivo in nondifferentiating cells, this would represent a decrease of $\leq 1.6\%$ in the total number of 5-mC residues in the genome. A change of this magnitude is below the level of significance for currently available methods which rely on chromatographic separation for quantitation of the overall 5-mC content of DNA, or on changes in the average size distribution of DNA fragments created by digestion of the total genomic DNA with methylation-sensitive restriction endonucleases such as *Hpa*II or *Hha*I. Loss of methyl groups from DNA induced by Me_2SO treatment of FL cells is not detected as a change in the ratio of 5-mC to $(C + 5\text{-mC})$ by HPLC analysis of bases in hydrolysates of FL cell DNA (CHRISTMAN et al. 1980) or as a change in the percentage of 5-mC in the penultimate position of CCGG sites by reversed-phase HPLC analysis of nucleotides derived from DNA labeled at the 5′ ends with ^{32}P after digestion with *Msp*I (J.K. Christman unpublished result). Nevertheless, a loss of 300 000 5-mC residues per haploid genome is three orders of magnitude greater than would be expected if all CpG dinucleotides in the α, β-major, and β-minor globin genes were to be converted from a completely methylated state to a completely unmethylated state. This suggests that terminal differentiation of FL cells involves a change in the extent of methylation of a number of genome regions not immediately associated with globin genes.

Although the characterization and precise localization of these sites is far from complete, it is clear (a) that the C residues in DNA of Me_2SO-treated FL cells which can be methylated in vitro are not confined to a specific fraction of the genome, and (b) that Me_2SO treatment does not affect methylation of *Hpa*II and *Hha*I sites in FL cell globin genes. More than 95% of radiolabeled methyl groups transferred in vitro from ^{3}H-methyl-S-AdoMet to DNA isolated from Me_2SO-treated cells are found in 5-mC residues. These radiolabeled 5-mC residues are distributed among pyrimidine isostichs with the same frequency as 5-mC residues resulting from in vivo methylation of DNA in untreated FL cells. Similarly, the fraction of total radiolabeled 5-mC residues present in active and inactive chromatin, in reiterated and single-copy DNA, in satellite and main band DNA prepared from DNA of Me_2SO-treated cells after in vitro methylation precisely reflects the proportion of total 5-mC residues occurring in comparable fractions prepared from DNA of untreated FL cells (CHRISTMAN 1979 and unpublished observations). The simplest interpretation of these results, assuming that FL cell DNA MTase acts primarily at hemimethylated CpG sites, is that chemical inducers such as Me_2SO decrease the rate of methylation of newly synthesized DNA. This would lead to persistence of hemimethylated sites created during semiconservative replication of DNA and, in nonsynchronized cell cultures, an equivalent probability of occurrence of hemimethylation at any site which was previously methylated on both strands. Mechanisms by which such randomly generated sites could lead to specific changes in the pattern of methylation are presented in Sect. 4.

It is unfortunate from the point of view of attempting to determine the relationship between loss of methyl groups and activation of genes during FL cell differentiation that no change occurs in the extent of methylation of *Hpa*II and *Hha*I sites in the globin genes of hexamethylene-bis-acetamide-treated FL cells (SHEFFREY et al. 1982) or Me$_2$SO-treated FL cells (WEICH 1982). Only a few of the CpG dinucleotides in the β-globin genes are in sites accessible to analysis with *Hpa*II and *Hha*I and these sites lie well away from the coding regions and proposed 5′ regulatory sequences of the genes. The β-globin gene is flanked by two *Hpa*II sites, one $\sim$1.7 kb 5′ to the 5′ end of the coding region, the other 0.3 kb 3′ to the 3′ end of the coding region adjacent to a *Hha*I site. Both sites at the 3′ end of the gene are fully methylated and remain so whether the cells are differentiated or not. The *Hpa*II site 5′ to the gene is unmethylated in a fraction of the cells, but this fraction does not increase with differentiation. A fully methylated *Hpa*II site is also present in the β-minor gene of both untreated and Me$_2$SO-treated FL cells.

The α-globin gene contains one *Hpa*II and two *Hha*I sites within the coding region and several others in both the 5′ and 3′ flanking regions. Analysis of methylation of *Hpa*II and *Hha*I sites in the sequence of the family of α-globin genes reveals a complex pattern, with some sites partially methylated and others fully methylated. However, no detectable loss of methylation occurs at any of these sites during Me$_2$SO-induced differentiation. SHEFFREY et al. (1982) have demonstrated that despite the lack of change in pattern of methylation of CpG sites in the globin genes during FL cell differentiation, a change occurs in the configuration of chromatin which renders a small region ($\sim$200 bp) starting 50 bp 5′ to the coding region of the β-globin gene hypersensitive to DNaseI digestion. A similar hypersensitive site was tentatively localized to a position immediately adjacent to the 5′ end of the coding region of the α-globin gene. This finding contrasts with the report of MCGHEE et al. (1981), which indicates a 200-bp region 5′ to the adult chicken β-globin gene with a high frequency of CpG sites which is both low in 5-mC content and nuclease-sensitive when the globin gene is active. Thus, if loss of methyl groups from DNA is part of the mechanism involved in triggering globin gene expression in FL cells, it must be assumed that the loss of methyl groups occurs at C residues which are not in *Hpa*II or *Hha*I sites, that the pattern of methylation which would allow transcription of globin genes is already established in untreated FL cells, and/or that changes in methylation which trigger globin gene expression occur at sites far removed from the globin sequences.

3 The Effects of 5-Azacytidine and 5-Aza-2′-deoxycytidine on FL Cells

3.2 Effects on Growth, Metabolism, and Accumulation of Hemoglobin

5-AzaCR and 5-AzaCdR act as inducers of FL cell differentiation in a concentration-dependent manner over a rather narrow range. Maximal induction of 12%–15% of cells occurs 5–7 days after a 24-h exposure to 1–2 μM 5-azaCR or 0.2–0.4 μM 5-azaCdR, and is not enhanced by increasing time of exposure or initial concentration of analogs, as both changes lead to decreased cell growth and viability. Thus, even though 5-azaCR and 5-azaCdR act at concentrations a thousand times lower than inducers such as the bis-acetamides, butyrate, or L-ethionine, these cytosine analogs are weak inducers of FL cell differentiation.

After 24-h exposure to 5-azaCR (1 μM) or 5-azaCdR (0.25 μM), the rate of DNA synthesis in FL cells is reduced by $\sim 80\%$, although rates of RNA and protein synthesis are not inhibited by more than 15%–20%. Once FL cells are washed free of the analogs, at least 24 h is required for restoration of normal rates of DNA synthesis. This results in a 24–48 h lag in culture growth, and may account for reduced clonogenicity of the cells (CREUSOT et al. 1982). However, the analogs do not have a generalized toxic effect on the cells which prevents them from accumulating hemoglobin. The percentage of cells induced to differentiate by Me$_2$SO at optimal concentrations is the same regardless of whether or not the cells have been pretreated with 5-azaCR or 5-azaCdR (although differentiated cells appear sooner in cultures of analog-treated cells), and treatment of cells with 5-azaCR prior to culture with suboptimal concentrations of either Me$_2$SO or L-ethionine leads to a percentage of differentiated cells which approximates that predicted by summing the effect of 5-azaCR and the second inducer.

3.2 Methylation of DNA

Although 5-azaCR and 5-azaCdR are weak inducers of FL cell differentiation, they have a more profound effect on DNA methylation in these cells than any other inducing agent we have tested, including L-ethionine. Hypomethylation of DNA is detectable not only as increased methyl acceptance in vitro, but as a decrease in the percentage of 5-mC residues in total DNA as well as in CpG dinucleotides of MspI and TaqI sites (Table 4). Loss of methyl groups from DNA is dependent both on concentration and time of exposure to the analogs (Table 3). Although hypomethylation occurs more rapidly and is more pronounced at concentrations above the optimal for differentiation, both analogs cause significant hypomethylation of DNA at concentrations where they are effective inducers.

When the effects of 5-azaCR on differentiation of C3H10T$^1/_2$ cells were first noted (CONSTANTINIDES et al. 1977), it was proposed (OLSON 1979) that the compound might act through an effect on DNA methylation. Incorporation of 5-azacytosine (5-azaC) (which presumably cannot be methylated

Table 4. Methylation of DNA in FL cells after termination of a 20-h exposure to 5-azaCR or 5-azaCdR

Time in culture after treatment (h)	In vitro methyl acceptance (pmol/100 μg)	% C residues methylated in		
		Total DNA synthesized during treatment	*Msp*I sites	*Taq*I sites
5-azaCdR (0.4 μ*M*)				
0	100	2.11	45.3	37.8
24	32	2.95	58.2	44.1
48	15	3.49	67.9	48.8
5-azaCR (1 μ*M*)				
0	53	3.09	ND	ND
24	32	3.38		
48	11	ND		
Control (no treatment)	12	3.62	67.9	47.9

[a] Percentage of radiolabeled C found as 5-mC in DNA of cells exposed to 6-^3H-uridine during the final 12 h of a 20-h exposure to 5-azaCR or 5-azaCdR as indicated. At 20 h the cells were washed free of both analog and radiolabel and incubated for an additional 24–48 h in the absence of analog and 6-^3H-uridine before isolation of DNA. Control cells were radiolabeled for 12 h and then incubated for 20 h in the absence of radiolabel before isolating DNA. Details of this determination and the measurement of in vitro methyl acceptance of DNAs are given in CREUSOT et al. (1982). Extent of methylation in the penultimate C residues of *Msp*I and *Taq*I sites was determined using a modification of the method of CEDAR et al. (1979) as described in CHRISTMAN (1981)

because of the substitution of a nitrogen for carbon at position 5 in the pyrimidine ring) into a methylation site during DNA replication would prevent maintenance methylation at that site and lead to creation of a fully unmethylated site in half the progeny after another round of DNA replication. This concept was supported by later experiments by JONES and TAYLOR (1980), which showed that 5-azaCdR which should be incorporated directly into DNA caused both differentiation and hypomethylation of C3H10T$^1/_2$ cells at lower concentrations than 5-azaCR. However, the theory that DNA methylation is inhibited because of an inherent inability of 5-azaC residues in DNA to serve as methyl acceptors was difficult to reconcile with JONES and TAYLOR's finding (1980) that substitution of 5-azaC for fewer than 5% of C residues during the first 24 h treatment caused a decrease of 85% in the methylation of newly synthesized DNA. Our work with FL cells (CREUSOT et al. 1981) suggested an alternative explanation: that incorporation of 5-azaC into DNA caused a rapid and virtually irreversible inactivation of DNA MTase, and that the extent to which DNA in 5-azaCR-treated cells became hypomethylated depended primarily on the amount of DNA synthesized in the absence of active DNA MTase.

We have shown that hypomethylation of FL cell DNA can be detected within 4 h of initiating treatment with 10 μ*M* 5-azaCR. During this period,

5-azaCR has no effect on the rate of synthesis of DNA, RNA, or protein and causes no loss of cell viability. Nevertheless, the level of active DNA MTase which can be extracted from the treated cells is reduced by $\geq 90\%$. Loss of enzyme activity is blocked when the cells are prevented from synthesizing DNA by addition of hydroxyurea a few minutes prior to exposing cells to 5-azaCR or 5-azaCdR (CREUSOT et al. 1982). Since (a) hydroxyurea does not prevent transport of 5-azaCR (F. Creusot, unpublished observation), (b) 5-azaCR and 5-azaCdR at concentrations as high as 10 mM have no effect on the activity of FL cell DNA MTase in vitro, and (c) inhibition of RNA synthesis with actinomycin D does not prevent enzyme loss in analog-treated cells, it appears that 5-azaC must be incorporated into DNA to mediate inactivation of DNA MTase. It should be noted that loss of DNA MTase activity during 5-azaCR treatment is not unique to FL cells and has been found to occur in Ehrlich ascites tumor cells isolated from their hosts 24 h after injection with 5-azaCR (TANAKA et al. 1980), in *E. coli* grown in the presence of 5-azaCR (FRIEDMAN 1979), and in human promyeloid cells exposed to 5-azaCR for 24 h (HL-60: MENDELSOHN et al. 1981; CHRISTMAN et al. 1983). Thus, continued synthesis of DNA in the absence of active DNA MTase is highly likely to represent a general mechanism for 5-azaCR-mediated loss of methyl groups from DNA. However, on this basis it would be predicted that the degree to which hypomethylated sites persist in the progeny of 5-azaCR-treated cells should also be determined by the balance between the rates at which DNA MTase activity and DNA synthesis are restored after cells have been washed free of 5-azaCR.

In FL cells, normal levels of DNA MTase are restored much more rapidly than normal rates of DNA synthesis (CREUSOT et al. 1982). As a result, within 24–48 h after terminating treatment with concentrations of 5-azaCR or 5-azaCdR which allow the cells to recover and differentiate, the DNA of treated cells can no longer be distinguished from DNA of untreated cells by its ability to accept methyl groups in vitro or by the percentage of 5-mC in CpG dinucleotides of *Msp*I and *Taq*I sites. Radiolabeled C residues incorporated into DNA during 5-azaCR and 5-azaCdR treatment are also methylated during this period (Table 4). Thus, the DNA is no longer rich in hemimethylated sites detectable by in vitro methylation, and most of the hemimethylated CCGG and TCGA sites created during treatment, are remethylated. The loss of hemimethylated sites can be attributed not only to synthesis of normally methylated DNA as the cells recover, but also to methylation of hypomethylated DNA synthesized during the period of 5-azaCR treatment. The implications of this "remethylation" of hypomethylated DNA with regard to the effects of 5-azaCR as an inducer of differentiation will be discussed in Sect. 4. However, it should be noted here that the data in Table 4 do not rule out the possibility that a small percentage of fully unmethylated sites are created during and persist after 5-azaCR treatment. We have, in fact, detected GCGC sites associated with the α-globin genes and the long terminal repeats of endogenous retrovirus genes which become unmethylated after 5-azaCR treatment of FL cells (WEICH 1982; J.K. Christman and L Berl, unpublished work).

3.3 The Mechanism of Inhibition of DNA Methyltransferase by 5-Azacytidine and 5-Aza-2'-deoxycytidine

Our findings (a) that loss of DNA MTase activity from 5-azaCR and 5-azaCdR-treated cells required that DNA be synthesized in the presence of the analogs, (b) that enzyme loss could occur within 4 h but that restoration of activity required 20–48 h, and (c) that even rigorous extraction methods did not increase the yield of active enzyme obtained from analog-treated cells[1] strongly suggested that a direct interaction between 5-azaC-substituted DNA and DNA MTase resulted in an essentially irreversible inactivation of the enzyme and that recovery of activity required synthesis of additional enzyme. HPLC analysis of nucleotides in DNA from FL cells incubated for 20 h in the presence of 1 μM 4-^{14}C-5-azaCR indicated that 5-azaC had been incorporated as a stable component of DNA, but that exposure to the analog at this concentration resulted in replacement of 0.3% or less of the C residues in DNA with 5-azaC. Since more than 80% of DNA MTase activity was lost during treatment, this meant that substitution of one in 300 C residues with 5-azaC was sufficient to inactivate all of the DNA MTase in an FL cell. However, despite its 5-azaC substitution, DNA synthesized during 5-azaCR treatment could be extensively methylated in vivo once DNA MTase levels in the cells increased and, under in vitro conditions, not only failed to inhibit DNA MTase but, as shown in Table 3, was actually methylated five to ten times more efficiently than DNA isolated from untreated cells.

These apparently contradictory findings can be explained in terms of the probability that DNA MTase will encounter a site which it can methylate before it interacts with a 5-azaC residue which can mediate its inactivation. Under normal conditions, the primary substrates for DNA MTase are hemimethylated sites in the small fraction of DNA which has just been replicated but not yet methylated. Thus during the first few hours of 5-azaCR treatment DNA MTase activity will be lost because of the enzyme's interaction with the only fraction of DNA in the cell which contains 5-azaC residues, newly synthesized DNA. As treatment is continued, the content of hemimethylated sites in the DNA will increase because DNA synthesized subsequent to enzyme inactivation will not be methylated. On the other hand, DNA synthesized in the later stages of analog treatment is likely to have a lower 5-azaC content than DNA synthesized at the beginning of treatment, since the effective concentration of analog to which the cells are exposed

---

1 Both 5-azaCR and 5-azaCdR are unstable in neutral aqueous solution (LIN et al. 1981; BEISLER 1978). HPLC analysis of culture medium which initially contained 1 μM 4-^{14}C-5-azaCR indicated that more than 70% of the compound was converted to degradation products during 4 h incubation with FL cells (J.K. Christman, unpublished observation)

Table 5. Effect of preincubation of DNA methyltransferase with 5-azaC-substituted DNA prior to initiation of assay by addition of S-AdoMet

DNA concentration during preincubation (15 min, 37° C)	Addition to assay	cpm/20 min, 37° C
5 µg 5-azaC DNA	None	71 000
10 µg 5-azaC DNA	None	19 500
15 µg 5-azaC DNA	None	9 000
15 µg 5-azaC DNA	15 µg L-ethionine DNA	14 050
15 µg L-ethionine DNA	None	175 000
None	15 µg L-ethionine DNA	184 000

All assay tubes contained 3 U FL cell DNA MTase (0.3 M NaCl extract) and 12 µM ^3H-methyl-S-AdoMet, sp. act. 17 Ci/mmol. Values given are the average of duplicate determinations in a typical assay. Radiolabel incorporated into acid-precipitable material in the absence of added DNA was 1250 cpm. The 5-azaC-substituted DNA contained approximately 1 5-azaC/100 C residues

diminishes rapidly[1]. As a result, during the period of recovery from analog treatment or under in vitro conditions, DNA MTase will be able to interact with DNA in which hemimethylated sites are present at higher concentrations than inhibitory 5-azaC-substituted sites. This argument is consistent with the finding that under the conditions used to assay methyl acceptance of DNAs, i.e., a high ratio of enzyme to substrate, the predominant reaction detected is methyl transfer rather than enzyme inactivation. However, when the ratio of 5-azaC-substituted DNA to enzyme is increased, incorporation of methyl groups into DNA does not reach a plateau level as would be expected if the enzyme were rate-limiting (and as is found with DNA from untreated or L-ethionine-treated cells), but instead falls off, indicating enzyme inhibition (CHRISTMAN et al. 1982; JONES and TAYLOR 1981).

The ability of 5-azaC-substituted DNA to inhibit DNA methyltransferase can be clearly demonstrated (Table 5) by preincubating the enzyme with DNA under conditions where methylation cannot occur (no AdoMet is added). The extent of inhibition of DNA MTase is dependent on the ratio of enzyme to 5-azaC-substituted DNA, and once inactivated the enzyme is no longer capable of methylating L-ethionine DNA added to the reaction mix. This suggests, as did our inability to extract active DNA MTase from nuclei of 5-azaCR-treated cells with 1 M NaCl or by digesting chromatin DNA with micrococcal nuclease and DNaseI (CREUSOT et al. 1982), that DNA MTase either becomes inactivated by an essentially irreversible bind-

---

1 The inability to recover active enzyme in 0.3 M NaCl extracts from 5azaCR- or 5azaCdR-treated cells is not a result of difficulties involved in isolating intact nuclei from such cells, leakage of enzyme from the nuclei prior to enzyme extraction or a newly developed resistance of chromatin proteins to extraction with 0.3 M NaCl [protein concentrations in extracts of treated and untreated cells do not vary from one another by more than 10% and contain identical levels of another chromatin associated enzyme, histone methyltransferase (R. Copp and J.K. Christman, unpublished observation)]

Table 6. Binding of FL cell DNA to protein extracted from FL cell nuclei with 0.3 M NaCl

Incubation for 20 min at 37° C with	cpm bound to filter	
	DNA from cells exposed to 10 μM 5-azaCR for 4 h	DNA from untreated cells
1. 0.5 U DNA methyltransferase, 0.6% sarkosyl, 0.5 M NaCl added prior to filtration	59 350	7 200
2. 0.5 U DNA methyltransferase and 0.6% sarkosyl added before incubation	850	900
3. No additions	1 950	2 250

DNA was isolated from cells incubated for 4 h with ^3H-methyl-thymidine in the presence or absence of 5-azaCR. To assay binding activity the DNA was incubated with 0.3 M NaCl extract of FL cell nuclei under conditions identical to those used to assay enzyme activity but without S-AdoMet. Each reaction contained 2.6 µg DNA (sp. act. 20×10^6 cpm/mg for DNA from untreated cells; 23×10^6 cpm/mg for DNA from 5-azaCR-treated cells; both DNA preparations were digested with S1 nuclease before use to reduce nonspecific binding to nitrocellulose filters) and 0.3 M NaCl extract containing 0.5 U DNA methyltransferase in 250 µl. After incubation, the mixtures were diluted to 1 ml with 10 mM Tris-HCl, pH 7.4, 0.1 mM EDTA, 0.5 M NaCl, and 0.6% sarkosyl and washed with 3 5-ml aliquots of the same buffer. Filters were dried and counted in nonaqueous fluor. If sarkosyl and NaCl were omitted from the diluting buffer and the wash buffer, 58 200 cpm 5-azaC DNA and 51 800 cpm DNA from untreated cells (100% input DNA) was bound to the filters

ing to 5-aza-C residues in DNA or binds normally to these residues and is destroyed during the process of attempting to transfer methyl groups to them. These results are analogous to those reported by FRIEDMAN (1981) for inhibition of bacterial DNA (C-5) MTases by DNA from *E. coli* K12 grown in the presence of 5-azaCR. However, it should be noted that DNA isolated from 5-aza-CR-treated *E. coli* differs from DNA isolated from 5-azaCR-treated mammalian cells (TANAKA et al. 1980; JONES and TAYLOR 1981; CREUSOT et al. 1982; CHRISTMAN et al. 1982) in that it is not a good substrate for in vitro methylation by homologous enzymes.

Although we cannot yet conclude that FL cell DNA MTase is inactivated by a direct interaction with 5-azaC residues in methylation sites, the data presented in Table 6 support the hypothesis that the enzyme is inactivated through a mechanism involving tight binding to DNA. DNA:protein complexes formed between proteins in 0.3 M NaCl extracts of nuclei and DNA from untreated FL cells can be disrupted by treatment with 0.5–1 M NaCl, while the complexes formed between proteins in these extracts and 5-azaC-substituted DNA withstand treatment with 1 M NaCl and 0.6% sarkosyl. DNA isolated from the sarkosyl-resistant complexes is considerably richer in 5-azaC than is total DNA from 5-azaCR-treated cells (J.K. Christman and G. Acs, unpublished observation), indicating that the 5-azaC residues themselves are important for formation of the complexes.

Characterization of the complexes is still is progress. However, the following observations have been made:

1. Complexes are disrupted by treatment with 1% SDS. This indicates that complex formation does not result from covalent linkages between protein and 5-azaC residues in DNA.

2. While complex formation is dependent on incubation time, temperature, and concentration of active DNA MTase, it does not require addition of AdoMet and is not inhibited by addition of 1 mM S-adenosylhomocysteine. This indicates that transfer of methyl groups to DNA is not required for complex formation.

3. Nuclear extracts from 5-azaCR-treated cells are depleted both of proteins capable of forming sarkosyl-resistant complexes with 5-azaC-substituted DNA and of DNA MTase activity.

4. The bulk of protein(s) responsible for formation of sarkosyl-resistant complexes with 5-azaC-substituted DNA have an affinity for binding to S-adenosylhomocysteine sepharose similar to that of DNA MTase (Fig. 1).

These results are consistent with a mechanism for DNA methyltransferase inactivation which involves an irreversible (under physiological conditions) binding of the enzyme to 5-azaC-rich regions of the DNA and suggest that DNA MTase is at least one of the proteins involved in the formation of sarkosyl-resistant 5-azaC DNA:protein complexes. However, it is evident from the data presented in Fig. 1 that proteins other than active DNA MTase are involved in complex formation. Fractions which have almost undetectable amounts of DNA MTase can still form complexes, and the ratio between complex-forming activity and DNA MTase activity is not constant in all fractions. It remains to be determined whether these proteins are inactive subunits of DNA MTase, regulatory proteins which recognize 5-azaC residues in methylation sites because they are involved in determining the specificity of methyl transfer, or proteins which are not involved in DNA methylation but which bind specifically to 5-azaC-substituted DNA because they recognize changes in DNA conformation induced by 5-azaC.

4 Conclusions and Speculations

The literature dealing with modification of C residues in DNA and its function in cells has a tradition of controversy which began with the first report of 5-mC in DNA (JOHNSON and COGHILL 1925). This finding was not substantiated by later investigators and probably resulted from use of a 5-mC standard which contained almost as much C as 5-mC (WYATT 1951; VISCHER et al. 1949). There was also a tradition of regarding these modifications as irrelevant, since they had no demonstrable function. This view has changed in recent years, but controversy continues, at least in part, because the methodology has not yet been developed for examining all methylation sites in a particular gene or gene region, a situation which has forced a great deal of extrapolation from examination of a limited number of methylation sites which may or may not be representative. The task of trying to prove that methylation of DNA plays a role in regulating gene activity has been further complicated by a lack of knowledge about whether or not all methyl groups in DNA of vertebrate cells serve the same function, a dearth of information about how the presence or absence of 5-mC residues in a specific DNA sequence affects their interaction with

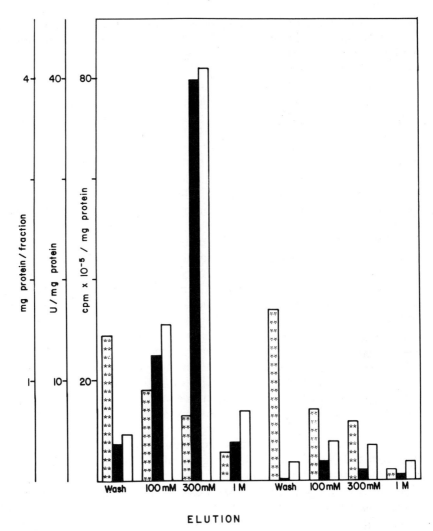

Fig. 1. Elution pattern of DNA methyltransferase and DNA (5-azaC) binding activity from *S*-adenosylhomocysteine Sepharose. The four sets of *bars on the left* show the elution pattern at the indicated molarity of NaCl for protein extracted from 5×10^8 nuclei of untreated FL cells with 0.3 *M* NaCl. The four sets of *bars on the right* show the elution pattern for protein extracted from 5×10^8 nuclei of FL cells exposed to 10 μ*M* 5-azaCR for 7 h. *Stippled bars* show total protein eluted in mg/fraction; *black bars* show DNA methyltransferase activity in U/mg protein; *open bars* show the capacity of proteins in the fraction to bind radiolabeled 5-azaC-substituted DNA in a complex resistant to 0.5 *M* NaCl and 0.6% sarkosyl. DNA methyltransferase activity was determined using acceptor DNA isolated from FL cells treated with L-ethionine. Sarkosyl-resistant DNA binding to nitrocellulose filters was measured as described in Table 6 using a DNA with sp. act. of 6×10^5 cpm/OD_{260}. Binding per milligram protein was calculated from the amount of protein which gave half-maximal binding of 60000 cpm of radiolabeled DNA. Protein content was determined using Bio-Rad Protein dye reagent

chromatin proteins or regulatory proteins and enzymes that act on DNA in chromatin, and a limited amount of direct evidence indicating how DNA methylation is regulated so that the requirement for specific loss or gain of methylated sites can be met.

The work reviewed here indicates that methylation inhibitors which can be administered in vivo may provide answers or at least clues as to how patterns of DNA methylation are established and as to how changes in these patterns affect gene function. Methylation inhibitors such as L-ethionine and 5-azaCR have already made it possible to obtain vertebrate DNAs with a sufficient degree of undermethylation to serve as methyl acceptors for use in in vitro studies on interaction of vertebrate DNA MTases with their natural substrates and to demonstrate the preference of vertebrate DNA MTases for ds DNAs with more methyl groups in one strand than in the other.

Some important aspects of the methylation process in vivo have also been revealed by studying the relationship between the time course of loss of DNA MTase activity, the rate of synthesis of DNA, and the extent of methylation of DNA in 5-azaCR- and 5-azaCdR-treated cells. Since 5-azaCR and 5-azaCdR must be incorporated into DNA to mediate inhibition of DNA synthesis, the finding that up to 85% of the DNA MTase activity in a culture of unsynchronized cells can be inhibited within 4 h (Table 3) suggests that within 4 h almost every molecule of DNA MTase in cells synthesizing DNA has interacted with newly synthesized 5-azaC-rich DNA. This result does not allow any conclusion to be made as two whether the enzyme normally moves in a processive manner along newly synthesized strands of DNA as suggested by DRAHOVSKY and MORRIS (1971), or whether it acts in a saltatory fashion, preferentially methylating newly synthesized DNA because of a high affinity for hemimethylated sites. However, the finding that DNA synthesized during 5-azaCR treatment can be extensively methylated many hours later once the cells are allowed to recover and DNA MTase levels are increased suggests that DNA MTase binds to and methylates hemimethylated sites in "old" nonreplicating DNA almost as efficiently as it binds to hemimethylated sites in newly replicated DNA. This indicates, as does the observation that hemimethylated DNA injected into oocytes becomes at least partially methylated in the absence of DNA replication (HARLAND 1982), that MTase is not an integral part of the replication complex itself and that it does not have to follow immediately behind the replication complex in order to act on hemimethylated sites.

Finally, the finding that L-ethionine and 5-azaCR can induce differentiation in a variety of cells serves as one of the primary arguments that loss of methyl groups from DNA may play a causal role in activating gene expression. However, the same types of contradictory observations which resulted from attempting to relate gene activity to patterns of methylation at restriction endonuclease sites (Sect. 1.2) have arisen during studies of the effects of DNA MTase inhibitors. These can be summarized as follows:

1. There is no direct relationship between the degree of hypomethylation of DNA caused by a particular inhibitor and the number of cells which will eventually go on to differentiate.

2. These inhibitors, which would a priori be expected to cause a random inhibition of methylation of C residues in newly synthesized DNA, do not randomly activate genes, but instead trigger expression of a different and limited repertoire of genes characteristic of the type of cell that is treated.
3. In some 5-azaCR-treated cells, changes in gene expression are detected prior to completion of two rounds of DNA replication.

It has been suggested that the failure of 5-azaCR to activate genes could indicate that some genes require more extensive demethylation than can be achieved without toxicity (FELSENFELD and McGHEE 1982). It is clear that both L-ethionine and 5-azaCR are toxic to cells and have a variety of effects in addition to inhibition of DNA MTase which could influence their ability to act as inducers of differentiation. However, under conditions employed to induce differentiation of FL cells (CHRISTMAN et al. 1977; CREUSOT et al. 1982), the failure of some cells to differentiate could not be ascribed to toxicity, since the treated cells were still capable of differentiating to response to Me$_2$SO. An alternative explanation, suggested by our finding that hemimethylated sites in 5-azaCR-treated cells can be remethylated once cells recover DNA MTase activity, is that only a limited number of sites in DNA which become hemimethylated during treatment with an inhibitor of DNA methylation remain unmethylated for a period long enough to allow another full round of DNA synthesis and establishment of a completely unmethylated site. If it is further assumed that only those genes in which completely unmethylated sites are established in regulatory regions are activated, some aspects of the effects of methylation inhibitors might be explained in terms of mechanisms by which fully unmethylated sites arise.

For example, a random escape of hemimethylated sites would predictably lead to the creation of a mixed population of cells, each with a different array of totally unmethylated sites. This mechanism is compatible with the low frequency (3×10^{-6}–10^{-3}) of gene activation observed in selective systems which require the expression of a previously repressed gene for cell survival (MOHANDAS et al. 1981; CLOUGH et al. 1982). However, random escape from methylation could not be expected to lead to appearance of a detectable number of cells with a specific change in pattern of methylation or capable of producing a specific gene product under nonselective conditions. This would seem to require a mechanism for blocking methylation of hemimethylated sites in specific DNA regions[1].

One feasible mechanism by which changes in the pattern of methylation could occur during normal development would be through interaction of DNA binding proteins (or other factors which lead to altered DNA conformation) with specific gene regions which become hemimethylated during DNA replication. If these factors prevented or slowed the rate of methylation, hemimethylated sites would persist in these regions and greatly increase

---

1 We have been able to demonstrate that chromatin proteins can limit the access of DNA MTase to sites which it normally methylates (CREUSOT and CHRISTMAN 1981). Whether this limitation is specific enough to indicate the existence of factors that regulate methylation of hemimethylated sites remains to be determined

the frequency with which specific completely unmethylated sites would occur after another round of cell division. Cell lines such as FL which continually generate small numbers of differentiated cells might be expected to have such factors already present in at least some individuals in the population. These factors would then account for the observed specificity with which inhibitors of methylation induce changes in gene expression in particular cell types under nonselective conditions because they would specifically prevent remethylation of hemimethylated sites created at random, leading to activation of the same genes as would be activated during normal differentiation. If, in fact, these factors were already bound to hemimethylated sites before the cells were exposed to methylation inhibitors, treatment might create fully unmethylated sites with only one round of DNA replication, triggering rapid alterations in gene expression. Characteristic variations in the numbers of cells containing hemimethylated sites and concentration of factors which stabilize them, as well as the speed with which levels of active DNA MTase are restored after 5-azaCR treatment, might also explain why 5-azaCR can cause hypomethylation and activation of up to 50% of EV-1 loci in chicken lymphocytes without affecting methylation of β-globin genes (GROUDINE et al. 1981), and why 5-azaCR can activate expression of hypoxanthine phosphoribosyl transferase genes or inactive human X chromosomes in mouse cells (MOHANDAS et al. 1981; LESTER et al. 1982) but not in diploid human cells (WOLF and MIDGEON 1982).

Another possibility which must be considered is that loss of a methyl group from only one strand of DNA creating a hemimethylated site in a critical gene region is sufficient to allow binding of regulatory proteins which allow gene expression under appropriate conditions, and that binding of regulatory proteins or changes in configuration as a result of expression prevent subsequent methylation at the hemimethylated site. This would be consistent with the finding of COMPERE and PALMITER (1981) that 5-azaCR treatment leads to activation of the metallothioneine-1 gene in cultured cells before one round of DNA replication can be completed. Evidence supporting the idea that hemimethylation at a critical site might allow gene activation has been obtained in studies of hormone-modulated developmental changes in gene expression. Activation of the chicken vitellogenin gene in response to estrogen treatment is associated with loss of methyl groups in *Hpa*II sites at the 5' end of the gene, but synthesis of vitellogenin can be detected before fully unmethylated sites appear (WILKS et al. 1982). The same fully unmethylated site appears in the vitellogenin gene in another estrogen-responsive tissue but does not result in gene expression.

Whether creation of hemimethylated sites can lead to gene activation and subsequent changes in specific gene regions which are fixed by complete loss of methylation at critical sites or whether complete loss of methylation at critical sites is required before gene activation occurs will have to remain a point of speculation until such sites have been identified and characterized. However, in terms of our current understanding of how methyl groups are lost from DNA[1] and how that loss correlates with altered gene expression, it is already clear that the relationship between the two events is not simple. It is probable that both processes will require the participation of a number of proteins or other factors with a high degree of tissue specificity. These would include factors which recognize specific hemimethylated sites and block their methylation, factors which bind to fully methylated sites to prevent access of proteins involved in transcription and, since it is obvious that loss of methyl groups may be necessary but is not in itself sufficient to activate gene expression, factors which would interact with either hemimethylated or fully unmethylated sites to regulate transcription in a tissue-specific manner. It is to be hoped that at least one outcome of studies of DNA methylation will be identification of tissue-specific regulatory factors through their affinity for synthetic or cloned sequences of DNA with completely unmethylated, hemimethylated, or fully methylated sites, so that complex processes involved in regulating gene activity during vertebrate development can finally be elucidated.

Acknowledgments. I would like to acknowledge the continuing valuable contributions of my colleague Dr. GEORGE ACS, as well as our fruitful associations with Dr. RICHARD COPP, Dr. FRANCINE CREUSOT, Ms. BETH SCHOENBRUN, and Dr. NADINE WEICH, whose doctoral research projects contributed so much to the work reported here. We are all indebted to Dr. P. PRICE for his help in annealing studies and for introducing us to the complexities of restriction endonuclease analysis, and to Mrs. NATALIE SCHNEIDERMAN for her superb technical assistance. Work from our laboratory described in this chapter was supported by UPHS Grants R01-CA16890.

References

Adams RLP (1974) Newly synthesized DNA is not methylated. Biochim Biophys Acta 335:365–373

Arber W, Linn S (1969) DNA modification and restriction. Annu Rev Biochem 38:467–500

Behe M, Felsenfeld G (1981) Effects of methylation on a synthetic polynucleotide: the B-Z transition in poly(dG-m5dC).poly(dG-m5dC). Proc Natl Acad Sci USA 78:1619–1623

Beisler JA (1978) Isolation, characterization and properties of a labile hydrolysis product of the antitumor nucleoside, 5-azacytidine. J Med Chem 21:204–208

Bird AP, Taggart MH, Smith BA (1979) Methylated and unmethylated DNA compartments in the sea urchin genome. Cell 17:889–901

Boehm TLJ, Drahovsky D (1981a) Hypomethylation of DNA in Raji cells after treatment with N-methyl-N-nitrosourea. Carcinogenesis 2:39–42

Boehm TLJ, Drahovsky D (1981b) Enzymatic hypomethylation of inverted repeats in transcribed DNA regions of mouse P815 mastocytoma cells. Int J Biochem 13:153–158

Boyer HW (1971) DNA restriction and modification mechanisms in bacteria. Annu Rev Microbiol 25:153–176

Brown FL, Musich PR, Maio JJ (1978) CaeI: an endonuclease isolated from the African monkey with properties indicating site-specific cleavage of homologous and heterologous mammalian DNA. Nucleic Acids Res 5:1093–1107

---

1 An alternate mechanism for loss of methyl groups which would not involve DNA replication is suggested by the finding that FL cell nucleoplasm contains an activity which can remove methylation in C residues of *Hpa*II sites. However, this activity does not increase during FL cell differentiation (GJERSET and MARTIN 1982)

74   J.K. Christman

Brunk CF, Jones KC, James TW (1979) Assay for nanogram quantities of DNA in cellular homogenates. Anal Biochem 92:497–500

Burdon RH, Adams RLP (1969) The *in vivo* methylation of DNA in mouse fibroblasts. Biochim Biophys Acta 174:322–329

Burdon RH, Martin BT, Lal BM (1967) Synthesis of low molecular weight ribonucleic acid in tumor cells. J Mol Biol 28:357–371

Cato AC, Burdon RH (1979) Mammalian DNA methylation and a nuclear *S*-adenosyl L-methionine-dependent nuclease activity. FEBS Lett 99:33–38

Cedar H, Solage A, Glaser G, Razin A (1979) Direct detection of methylated cytosine in DNA by use of the restriction enzyme MspI. Nucleic Acids Res 6:2125–2132

Chargaff E (1955) Isolation and composition of the deoxypentose nucleic acids and the corresponding nucleoproteins. In: Chargaff E and Davidson JN, Eds The nucleic acids, Vol I, chapter 10. Academic, New York, pp 307–371

Chargaff E, Crampton CF, Lipshitz R (1953) Separation of calf thymus deoxyribonucleic acid into fractions of different composition. Nature 172:289–292

Christman JK (1979) DNA methylation in Friend erythroleukemia cells. In: E. Usdin, R.T. Borchardt, C.R. Creveling Eds Transmethylation. Elsevier, New York, pp 493–502

Christman JK (1982) Separation of major and minor deoxyribonucleoside monophosphates by reverse-phase high-performance liquid chromatography: a simple method applicable to quantitation of methylated nucleotides in DNA. Anal Biochem 119:38–48

Christman JK, Price P, Pedrinan L, Acs G (1977) Correlation between hypomethylation of DNA and expression of globin genes in Friend erythroleukemia cells. Eur J Biochem 81:53–61

Christman JK, Weich N, Schoenbrun B, Schneiderman N, Acs G (1980) Hypomethylation of DNA during differentiation of Friend erythroleukemia cells. J Cell Biol 86:366–370

Christman JK, Creusot F, Acs G (1982) Mechanism of inhibition of DNA methyltransferase by 5-azacytosine substituted DNA. In: Biochemistry of *S*-adenosylmethionine and related compounds. Macmillan London, pp 223–229

Christman JK, Mendelsohn N, Herzog D, Schneiderman N (1983) Effect of 5-azacydtidine on differentiation and DNA methylation in human promyelocytic leukemia cells (HL-60). Cancer Res 43:763–769

Clough DW, Kunkel LM, Davidson RL (1982) 5-Azacytidine-induced reactivation of a herpes simplex thymidine kinase gene. Science 216:67–74

Compere SJ, Palmiter RD (1981) DNA methylation controls the inducibility of the mouse metallothionein-1 gene in lymphoid cells. Cell 25:233–240

Constantinides PG, Jones PA, Gevers W (1977) Functional striated muscle cells from non-myoblast precursors following 5-azacytidine treatment. Nature 274:775–780

Copp RP (1981) Studies of nuclear protein methyltransferase activity in Friend erythroleukemia cell Doctoral thesis, Faculty of the Graduate School of Arts and Science New York University

Coulandre C, Miller JH, Farabaugh PJ, Gilbert W (1978) Molecular basis of base substitution hotspots in *Escherichia coli*. Nature 274:364–366

Creusot F, Christman JK (1981) Localization of DNA methyltransferase in the chromatin of Friend erythroleukemia cells. Nucleic Acids Res 9:5359–5381

Creusot F, Herzog D, Acs G, Christman JK (1981) Inhibition of DNA methyltransferase and induction of Friend erythroleukemia cell differentiation by 5-azacytidine. Proc Am Assoc Cancer Res 22:48

Creusot F, Acs G, Christman JK (1982) Inhibition of DNA methyltransferase and induction of Friend erythroleukemia cell differentiation by 5-azacytidine and 5-aza-2'-deoxycytidine. J Biol Chem 257:2041–2048

Dimitriadis GJ, Tata JR (1980) Subnuclear fractionation by mild micrococcal nuclease treatment of nuclei of different transcriptional activity causes a partition of expressed and non-expressed genes. Biochem J 187:467–477

Doerfler W (1981) DNA methylation – a regulatory signal in eukaryotic gene expression. J Gen Virol 57:1–20

Doskocil J, Sôrm F (1962) Distribution of 5-methylcytosine in pyrimidine sequences of deoxyribonucleic acids. Biochim Biophys Acta 55:953–959

Drahovsky D, Morris NR (1971) Mechanism of action of rat liver DNA methyltransferase. II. Interaction with single-stranded methyl-acceptor DNA. J Mol Biol 61:343–356

Ehrlich M, Wang RYH (1981) 5-Methylcytosine in eukaryotic DNA. Science 212:1350–1357

Ehrlich M, Ehrlich K, Mayo JA (1975) Unusual properties of the DNA from *Xanthomonas* phage XP-12 in which 5-methylcytosine completely replaces cytosine. Biochim Biophys Acta 395:109–119

Ehrlich M, Gama-Sosa MA, Huang L-H, Midgett RM, Kuo KC, McCune RA, Gehrke CR (1982) Amount and distribution of 5-methylcytosine in human DNA from different types of tissues or cells. Nucleic Acids Res 10:2709–2721

Evans HH, Evans TE (1970) Methylation of the deoxyribonucleic acid of *Physarum polycephalum* at various periods during the mitotic cycle. J Biol Chem 245:6436–6441

Felsenfeld G, McGhee J (1982) Methylation and gene control. Nature 296:602–603

Fisher EF, Caruthers MH (1979) Studies on gene control regions XII. The functional significance of a *lac* operator constitutive mutation. Nucleic Acids Res 7:401–416

Fleissner E, Borek E (1962) Studies on the enzymatic methylation of soluble RNA. I. Methylation of the s-RNA-polymer. Biochem J 2:1093–1100

Fradin A, Manley J, Prives C (1982) Methylation of simian virus 40 HpaII sites affects late but not early viral gene expression. Proc Natl Acad Sci USA 79:5142–5146

Friedman S (1979) The effect of 5-azacytidine on *E. coli* DNA methylase. Biochem Biophys Res Commun 89:1328–1333

Friedman S (1981) Inhibition of DNA (cytosine-5) methylase by 5-azacytidine. The effect of azacytosine-containing DNA. Mol Pharmacol 19:314–320

Gantt R, Montes de Oca F, Evans VJ (1973) Methylation of satellite deoxyribonucleic acid in mouse neoplastic and non-neoplastic cell cultures. In Vitro 8:288–294

Garel A, Axel R (1976) Selective digestion of transcriptionally active ovalbumin genes from oviduct nuclei. Proc Natl Acad Sci USA 73:3966–3970

Gill JE, Mazrimus JA, Bishop CC (1974) Physical studies on synthetic DNAs containing 5-methylcytosine. Biochim Biophys Acta 335:330–348

Gjerset RA, Martin DW Jr (1982) Presence of a DNA demethylating activity in the nucleus of murine erythroleukemic cells. J Biol Chem 257:8581–8583

Glickman B, Van Den Elsen P, Radman M (1978) Induced mutagenesis in dam⁻ mutants of *Escherichia coli*: a role for 6-methyladenine residues in mutation avoidance. Mol Gen Genet 163:307–312

Gold M, Hurwitz J (1963) The enzymatic methylation of the nucleic acids. Cold Spring Harbor Symp Quant Biol 28:149–156

Gottesfeld JM, Butler (1977) Structure of transcriptionally active chromatin subunits. Nucleic Acids Res 4:3155–3173

Groudine M, Weintraub H (1982) Propagation of globin DNAseI hypersensitive sites in absence of factors required for induction: a possible mechanism for determination. Cell 30:131–139

Groudine M, Eisenman R, Weintraub H (1981) Chromatin structure of endogenous retroviral genomes and activation by an inhibitor of DNA methylation. Nature 292:311–317

Gruenbaum Y, Cedar H, Razin A (1982) Substrate and sequence specificity of a eukaryotic DNA methylase. Nature 295:620–622

Harbers K, Harbers B, Spencer JH (1984) Nucleotide clusters in deoxyribonucleic acids. X. Sequences of the pyrimidine oligonucleotides of mouse L-cell satellite DNA. Biochem Biophys Res Commun 58:814–821

Harbers K, Schneike A, Stuhlman H, Jahner D, Jaenisch R (1981) DNA methylation and gene expression: endogenous retroviral genome becomes infectious after molecular cloning. Proc Natl Acad Sci USA 78:7609–7613

Harland RM (1982) Inheritance of DNA methylation in microinjected eggs of *Xenopus laevis*. Proc Natl Acad Sci USA 79:2323–2327

Holliday R, Pugh JE (1975) DNA modification mechanisms and gene activity during development. Science 187:226–232

Johnson TB, Coghill RD (1925) Researches on pyrimidines. CIII. The discovery of 5-methylcytosine in tuberculinic acid, the nucleic acid of the tubercle bacillus. J Am Chem Soc 47:2838–2844

Jones PA, Taylor SM (1980) Cellular differentiation, cytidine analogs and DNA methylation. Cell 20:85–93

Jones PA, Taylor SM (1981) Hemimethylated duplex DNAs prepared from 5-azacytidine-treated cells. Nucleic Acids Res 9:2933–2947

Kappler JW (1971) The 5-methylcytosine content of DNA: tissue specificity. J Cell Physiol 78:33–36

Kaput J, Sneider JW (1979) Methylation of somatic vs germ cell DNAs analyzed by restriction endonuclease digestion. Nucleic Acids Res 7:2303–2322

Korba BE, Hays JB (1982) Partially defient methylation of cytosine in DNA at CC(A/T)GG sites stimulates genetic recombination of bacteriophage lambda. Cell 28:531–541

Kornberg A, Zimmerman SB, Kornberg SR, Josse J (1959) Enzymatic synthesis of deoxyribonucleic acid. Influence of bacteriophage T2 on the synthetic pathway in host cells. Proc Natl Acad Sci USA 45:772–785

Lapeyre JN, Becker FF (1979) 5-Methylcytosine content of nuclear DNA during chemical hepatocarcinogenesis and in hepatocarcinomas which result. Biochem Biophys Res Commun 87:698–705

Lapeyre JN, Walker MS, Becker FF (1981) DNA methylation and methylase levels in normal and malignant mouse hepatic tissues. Carcinogenesis 2:873–878

Lester SM, Korn NJ, DeMars R (1982) Derepression of genes on the human inactive X chromosome: evidence for differences in locus-specific rates of transfer of active and inactive genes after DNA-mediated transformation. Somatic Cell Genet 8:265–284

Lieberman MW (1973) Fractionation of mouse DNA in preparative $AG + -Cs_2SO_4$ gradients. Biochim Biophys Acta 324:2193–2197

Lin KT, Momparler RL, Rivard GE (1981) High performance liquid chromatography analysis of chemical stability of 5-aza-2'-deoxycytidine. J Pharm Sci 70:1228–1232

Lu L-J, Randerath K (1980) Mechanism of 5-azacytidine induced tRNA:cytosine-5-methyltransferase deficiency. Cancer Res 22:2701–2705

Mandel JL, Chambon P (1979) DNA methylation: organ specific variations in the methylation pattern within and around ovalabumin and other chicken genes. Nucleic Acids Res 7:2081–2103

Marinus MG, Morris NR (1975) Pleiotropic effects of a DNA adenine methylation mutation (dam-3) in Escherichia coli K12. Mutat Res 28:15–26

Marks P, Rifkind (1978) Erythroleukemic differentiation. Annu Rev Biochem 47:419–448

McGhee JD, Ginder GD (1979) Specific DNA methylation sites in the vicinity of the chicken β-lobin gene. Nature 280:419–420

McGhee JD, Wood WI, Dolan M, Engel JD, Felsenfeld G (1981) A 200 base pair region at the 5' end of the chicken adult β-globin gene is accessible to nuclease digestion. Cell 27:45–55

Mendelsohn N, Michl J, Gilbert HS, Acs G, Christman J (1980) L-Ethionine as an inducer of differentiation in human promyelocytic leukemia cells (HL-60). Cancer Res 40:3206–3210

Mendelsohn N, Herzog D, Christman JK (1981) 5-Azacytidine as an inducer of differentiation in human leukemia cells (HL-60). Proc Am Assoc Cancer Res 22:48

Mohandas T, Sparkes RS, Shapiro LJ (1981) Reactivation of an inactive human X chromosome: evidence for X inactivation by DNA methylation. Science 211:393–396

Ohta TM, Tanaka M, Terada M, Miller O, Banks A, Marks P, Rifkind R (1976) Erythroid cell differentiation: murine erythroleukemia cell variant with unique pattern of induction by polar compounds. Proc Natl Acad Sci USA 73:1232–1236

Olson CB (1979) 5-Methylcytosine, 5-azacytidine and development: a synthesis. Speculation in Science and Technology 2:365–373

Orkin SR, Harosi FI, Leder P (1979) Differentiation in erythroleukemia cells and their somatic hybrids. Proc Natl Acad Sci USA 72:98–102

Pages M, Roizes G (1982) Tissue specificity and organization of CpG methylation in calf satellite DNA I. Nucleic Acids Res 10:565–576

Pfohl-Leszkowicz A, Salas CE, Fuchs R, Dirheimer G (1981) Inhibition of DNA methylation by DNA-AAF. Biochemistry 20:3020–3024

Pollack Y, Swihart M, Taylor JH (1978) Methylation of DNA in early development: 5-methylcytosine content of DNA in sea urchin sperm and embryos. Nucleic Acids Res 5:4855–4863

Pollack Y, Stein R, Razin A, Cedar H (1980) Methylation of foreign DNA sequences in eukaryotic cells. Proc Natl Acad Sci USA 77:6463–6467

Preisler HD, Housman D, Scher W, Friend C (1973) Effects of 5-bromo-2-deoxyuridine on production of globin messenger RNA in dimethylsulfoxide-stimulated Friend erythroleukemia cells. Proc Natl Acad Sci USA 70:2956–2959

Razin A, Riggs AD (1980) DNA methylation and gene function. Science 210:604–610

Riggs AD (1975) X inactivation, differentiation and DNA methylation. Cytogenet Cell Genet 14:9–25

Rovera G, O'Brien TA, Diamond L (1977) Tumor promoters inhibit spontaneous differentiation of Friend erythroleukemia cells in culture. Proc Natl Acad Sci USA 74:2894–2898

Roy PH, Weissbach A (1975) DNA methylase from HeLa cell nuclei. Nucleic Acids Res 2:1669–1684

Sager R, Kitchin R (1975) Selective silencing of eukaryotic DNA. Science 189:426–433

Salomon R, Kaye AM, Herzberg M (1969) J Mol Biol 43:581–592 Mouse nuclear satellite DNA: 5-methylcytosine content, pyrimidine isoplith distribution and electron microscopic appearance

Scarano E, Iaccarino I, Grippo P, Parisi E (1967) The heterogeneity of thymine methyl group origin in DNA pyrimidine isostichs. Proc Natl Acad Sci USA 57:1394–1400

Seale RL (1978) Nucleosomes associated with newly replicated DNA have an altered conformation. Proc Natl Acad Sci USA 75:2717–2721

Sheffery M, Rifkind R, Marks P (1982) Murine erythroleukemia cell differentiation: DNase I hypersensitivity and DNA methylation near the globin genes. Proc Natl Acad Sci USA 79:1180–1184

Sheid B, Srinivasan PR, Borek E (1968) Deoxyribonucleic acid methylase of mammalian tissue. Biochem J 7:280–285

Simon D, Grunert F, v. Acken U, Doring HP, Kroger H (1978) DNA methylase from regenerating rat liver: purification and characterization. Nucleic Acids Res 5:2153–2167

Simon D, Tischer I, Wagner H, Werner E, Kroger H (1982) Effect of DNA methylation in vitro on the expression of SV-40 and VSPV. In: E. Usdin, R.T. Borchardt, C.R. Oreveling, Eds Biochemistry of S-adenosylmethionine and related compounds. Macmillan, London, pp 267–273

Singer J, Roberts-Ems J, Luthardt FW, Riggs AD (1979) Methylation of DNA in mouse early embryos, teratocarcinoma cells and adult tissues of mouse and rabbit. Nucleic Acids Res 7:2369–2385

Sinsheimer RL (1954) The action of pancreatic desoxyribonuclease. I. Isolation of mono- and dinucleotides. J Biol Chem 208:445–459

Smith SS, Yu JC, Chen CW (1982) Different levels of DNA modification at 5′CCGG in murine erythroleukemia cells and the tissues of normal mouse spleen. Nucleic Acids Res 10:4305–4319

Sneider TW (1980) The 5′ cytosine in CCGG is methylated in two eukaryotic DNAs and MspI is sensitive to methylation at this site. Nucleic Acids Res 8:3829–3840

Sneider TW, Potter VR (1969) Methylation of mammalian DNA: studies on Novikoff hepatoma cells in tissue culture. J Mol Biol 271–284

Sneider TW, Teague WM, Rogachevsky LM (1975) S-Adenosylmethionine:DNA-cytosine 5-methyltransferase from a Novikoff rat hepatoma cell line. Nucleic Acids Res 2:1685–1700

Sneider TW, Kaput J, Neiman D, Westmoreland B (1979) Hemimethylation of DNA: a basis for genetic recombination? In: E Usdin, RT Borchardt, CR Creveling (Eds) Transmethylation. Elsevier, New York, pp 473–481

Srinivasan PR, Borek E (1964) Enzymatic alteration of nucleic acid structure. Science 145:548–553

Stein R, Greunbaum Y, Pollack Y, Razin A, Cedar H (1982a) Clonal inheritance of the pattern of DNA methylation in mouse cells. Proc Natl Acad Sci USA 79:61–65

Stein R, Razin A, Cedar H (1982b) In vitro methylation of the hamster adenine phosphoribosyltransferase gene inhibits its expression in mouse L cells. Proc Natl Acad Sci USA 79:3418–3422

Sturm K, Taylor JH (1981) Distribution of 5-methylcytosine in DNA of somatic and germline cells from bovine tissues. Nucleic Acids Res 9:4537–4546

Stucliffe JG (1979) Complete nucleotide sequence of the *E. coli* plasmid pBR322. Cold Spring Harbor Symp Quant Biol 43:77–89

Tanaka M, Hibasami H, Nagai J, Ikeda T (1980) Effect of 5-azacytidine on DNA methylation in Ehrlich's ascites tumor cells. Aust J Exp Biol Med Sci 558:391–396

Turnbull JF, Adams RLP (1976) DNA methylase: purification from ascites cells and the effects of various DNA substrates on its activity. Nucleic Acids Res 3:677–695

Vanyushin BF, Tkacheva SG, Belozersky AN (1970) Rare bases in animal DNA. Nature 225:948–949

Vanyushin BF, Mazin AL, Vasilyev VK, Belozersky AN (1973) The content of 5-methylcytosine in animal DNA: the species and tissue specificity. Biochim Biophys Acta 229:397–403

Vardimon L, Kressmann A, Cedar H, Maechler M, Doerfler W (1982) Expression of a cloned adenovirus gene is inhibited by *in vitro* methylation. Proc Natl Acad Sci USA 79:1073–1077

Vischer E, Zamenhoff S, Chargaff E (1949) Microbial nucleic acids: the desoxypentose nucleic acids of avian tubercle bacilli and yeast. J Biol Chem 177:429–438

Waalwijk C, Flavell RA (1978) DNA methylation of a CCGG sequence in the large intron of the rabbit *β*-globin gene: tissue-specific variation. Nucleic Acids Res 5:4631–4641

Waechter DE, Baserga R (1982) Effect of methylation on expression of microinjected genes. Proc Natl Acad Sci USA 79:1106–1110

Watson JD, Crick FHC (1953) Molecular structure of nucleic acid. A structure for deoxyribose nucleic acid. Nature 171:964–967

Weich N (1982) DNA methylation in differentiating mouse erythroleukemia cells. Doctoral thesis, Graduate Faculty in Biomedical Studies, City University of New York

Weintraub H, Groudine M (1976) Chromosomal subunits in active genes have altered conformation. Science 193:848–856

Wigler M, Levy D, Perucho M (1981) The somatic replication of DNA methylation. Cell 24:33–40

Wilks AF, Cozens PJ, Mattai IW, Jost P-J (1982) Estrogen induces a demethylation at the 5′ end region of the chicken vitellogenin gene. Proc Natl Acad Sci USA 79:4252–4255

Wolf SF, Midgeon BR (1982) Studies of X chromosome DNA methylation in normal human cells. Nature 295:667–671

Wyatt GR (1951) Recognition and estimation of 5-methylcytosine in nucleic acids. Biochem J 48:581–584

Yamasaki H, Fibach E, Nudel U, Weinstein IB, Rifkind RA, Marks P (1977) Tumor promoters inhibit spontaneous and induced differentiation of murine erythroleukemia cells in culture. Proc Natl Acad Sci USA 74:3451–3455

Youssoufian H, Hammer SM, Hirsch M, Mulder C (1982) Methylation of the viral genome in an in vitro model of herpes simplex virus latency. Proc Natl Acad Sci USA 79:2207–2210

Yuan R, Meselson M (1970) A specific complex is formed between a restriction endonuclease and its DNA substrate. Proc Natl Acad Sci USA 65:357–362

DNA Methylation and Its Functional Significance: Studies on the Adenovirus System

W. Doerfler

1 Introduction

As in many problems related to the molecular biology of eukaryotic cells, virus systems have played an important role in elucidating the functional significance of DNA methylation. From many different lines of evidence gleaned from viral and nonviral systems, the notion has evolved that DNA methylation at specific sites, which may be different in individual genes or groups of genes, somehow has a function in the regulation of gene expression. At this time, we are far from understanding the biochemical mechanisms by which DNA modifications exert this regulatory effect. It is likely, though not proven, that DNA modifications, notably 5-methylcytosine (5mC) in the DNA of higher eukaryotes and 5mC and N^6-methyladenine (6mA) in the DNA of lower eukaryotes and prokaryotes, affect DNA-protein interactions in either positive or negative ways. The interference

Institute of Genetics, University of Cologne, Weyertal 121, D-5000 Köln 41

Current Topics in Microbiology and Immunology, Vol. 108
©Springer-Verlag Berlin · Heidelberg 1984

of methylated DNA sequences with the activity of restriction endonucleases constitutes the most clearly documented example to date of an effect of DNA modification on DNA-protein interactions. In the modulations of these interactions, methylated bases in DNA might constitute signals in their own right, or might assume signal values in fundamentally altering the structure of DNA by causing transitions from the conventional B form to the Z form (BEHE and FELSENFELD 1981) or from the B form to the A form of DNA (CONNER et al. 1982). At present, it is impossible to distinguish between these alternatives.

With the presumptive role of DNA methylation in eukaryotic gene expression, a stage has been reached in our investigations and in our interpretations where most, though not all, of the data seem to support a plausible model. The presence of 5mC in specific sequences, frequently at the 5′-CCGG-3′ sites and predominantly in the 5′ and promoter parts of specific genes (KRUCZEK and DOERFLER 1982, 1983; FLAVELL et al. 1982; OTT et al. 1982; STEIN et al. 1983; BUSSLINGER et al. 1983), causes transcriptional inactivation of these genes. Are the data emanating from many different eukaryotic systems persuasive enough for us to accept these conclusions? What are the mechanisms relating DNA methylation and gene regulation? What alternative or additional functions might DNA methylation affect? It will be important to consider these questions critically in the planning of future research in this rapidly advancing field.

Regulation of gene expression in eukaryotes is one of the fascinating topics in molecular biology, with interesting implications for differentiation and many areas of biology, e.g., virology, immunology, genetic disease, and oncology. The function that methylated bases in DNA fulfill seems to depend strongly on the base involved and on its highly specific position within, or in the vicinity of, regulatory sequences of genes. One of the major challenges in research in molecular biology is the elucidation of more complex coding principles which play important roles as regulatory signals in DNA. It is probably too simplistic to search for these signals exclusively in specific DNA sequences; they may also be found in complex structures of DNA or DNA-protein complexes. In recent years, it has become increasingly apparent that the double-stranded DNA molecule can assume a number of structures which differ strikingly from the classical B-form of DNA (for recent reviews, see *Cold Spring Harbor Symposium on Quantitative Biology* 1982, vol 47; DICKERSON et al. 1982; ZIMMERMAN 1982). It is uncertain at present to what extent these structures of DNA occur naturally or whether and how they can be implicated in special functions of DNA.

In this brief review I shall summarize the results of experiments that have employed the adenovirus system in an attempt to elucidate the functional roles that DNA methylation can play. DNA methylation and its functional implications have been covered extensively in a series of review articles (RAZIN and RIGGS 1980; BURDON and ADAMS 1980; DOERFLER 1981; EHRLICH and WANG 1981; HATTMAN 1981; RAZIN and FRIEDMAN 1981; DOERFLER 1983; COOPER 1983).

2 DNA Methylation and the Control of Eukaryotic Gene Expression

The regulation of gene expression in eukaryotes is most likely to involve highly specific DNA-protein interactions. Proteins binding to specific sequences in eukaryotes are only now being recognized (e.g., JACK et al. 1982), and their regulatory function is still uncertain. DNA methylation at specific sequences may modulate these interactions; DNA methylation of C residues in specific sequences is usually associated with the inactive state of genes.

Hypotheses relating DNA methylation and gene activity evolved from observations on the differential states of DNA methylation in developing organisms and in different tissues (RIGGS 1975; SAGER and KITCHIN 1975; HOLLIDAY and PUGH 1975; CHRISTMAN et al. 1977; BIRD 1978; MCGHEE and GINDER 1979; MCGHEE and FELSENFELD 1980). In a given gene that was actively expressed in one organ, all or most of the 5'-CCGG-3' sites were found to be unmethylated, while the DNA in the same gene in other organs not expressing this gene was completely methylated at the 5'-CCGG-3' sites. These sites were preferentially investigated because of the availability of the isoschizomeric restriction endonuclease pair *Hpa*II and *Msp*I (WAAL-WIJK and FLAVELL 1978), which permitted determination of the state of methylation at all 5'-CCGG-3' sites. In many genes the state of methylation at these sites appears to have special significance. In this way, inverse correlations between the extent of DNA methylation at 5'-CCGG-3' sites in specific genes and the degree to which these genes are expressed have been established in several different systems (SUTTER and DOERFLER 1979; DESRO-SIERS et al. 1979; MANDEL and CHAMBON 1979; SUTTER and DOERFLER 1980; VARDIMON et al. 1980; COHEN 1980; GUNTAKA et al. 1980; VAN DER PLOEG and FLAVELL 1980; VAN DER PLOEG et al. 1980; WEINTRAUB et al. 1981; KUHLMANN and DOERFLER 1982; KRUCZEK and DOERFLER 1982; SMITH et al. 1982). In some cases, these inverse correlations have been extended to other sequences containing 5'-CpG-3' dinucleotides, e.g., the sequence 5'-GCGC-3', the recognition sequence of the restriction endonuclease *Hha*I (KUHL-MANN and DOERFLER 1982; KRUCZEK and DOERFLER 1982). An inverse relationship between the extents of DNA methylation and gene expression could obviously indicate cause or consequence of gene inactivation. Investigations directed toward the core of the problem used in vitro-methylated, cloned genes that were microinjected into the nucleus of oocytes of *Xenopus laevis* (VARDIMON et al. 1981, 1982a, b; FRADIN et al. 1982) or into mammalian cells in culture (WAECHTER and BASERGA 1982; STEIN et al. 1982). Genes methylated at specific sites were not usually expressed in these experiments. The use of sequence-specific DNA methyltransferases from prokaryotic organisms allowed the determination of highly specific sequences, presumably involved in gene regulation (VARDIMON et al. 1981, 1982a, b; WAECHTER and BASERGA 1982; STEIN et al. 1982; FRADIN et al. 1982). Other sequences, when methylated, did not exhibit any effect on gene expression (DOERFLER et al. 1982; VARDIMON et al. 1982b). These data, accumulating from work in different systems and in different laboratories provided strong, direct evidence for the notion that methylated sequences at highly specific sites and sequences play an important part in the regulation of gene expression.

Further evidence comes from work with the cytidine analogue, 5'-azacytidine. This compound can be incorporated into replicating DNA (CHRISTMAN et al. 1980; JONES and TAYLOR 1981; CREUSOT et al. 1982; JONES et al. 1982), and – due to its chemical structure (N in the 5 position, instead of C) – cannot be methylated. However, the main inhibitory role of 5-azacytidine toward DNA methylation appears to emanate from its ability to inhibit DNA methyltransferases (CHRISTMAN et al. 1980; JONES and TAYLOR 1981; CREUSOT et al. 1982; JONES et al. 1982). In a number of different systems it has been possible to activate previously dormant genes and induce their expression (TAYLOR and JONES 1979; GROUDINE et al. 1981; JONES et al. 1982; HARRIS 1982). The original report on the gene-activating property of 5-azacytidine described the activation of a complex set of cellular functions, leading to in vitro differentiation of mouse fibroblasts to twitching muscle cells (CONSTANTINIDES et al. 1977). It is not to be expected that all dormant cellular genes can be activated by treatment of cells with 5-azacytidine. Absence of DNA methylation appears a necessary, but not in itself sufficient, precondition for gene activation (VAN DER PLOEG and FLAVELL 1980; KUHLMANN and DOERFLER 1982; OTT et al. 1982). A crucial mechanism such as gene activity may be subject to multifaceted regulatory mechanisms, DNA methylation constituting one important parameter for long-term gene inactivation. Thus, depending on the stringency of inactivation for a given gene, 5-azacytidine treatment may or may not lead to the activation of a certain gene or set of genes.

For a complete understanding of the role of DNA methylation in gene regulation, methods will be required to determine the state of methylation at all the 5'-CpG-3' sites in a gene and its adjacent sequences. Improved in vitro transcription systems will also be required to evaluate the effect of methylation in different 5'-CpG-3' sites. Presently available in vitro transcription systems do not seem to respond to DNA methylation.

3   Virion DNA and Free Viral DNA in Infected Cells Is Not Detectably Methylated

At present there is no evidence that virion DNA extracted from purified adenovirus particles, or free intracellular viral DNA extracted from productively or abortively infected cells, is detectably methylated (GÜNTHERT et al. 1976; SUTTER et al. 1978; VON ACKEN et al. 1979; SUTTER and DOERFLER 1979, 1980; VARDIMON et al. 1980; EICK et al. 1983). This statement is supported by the results of experiments employing two-dimensional thin-layer chromatography (GÜNTHERT et al. 1976), differential cleavage of viral DNA (SUTTER et al. 1978; SUTTER and DOERFLER 1979, 1980) with the restriction endonuclease pair HpaII and MspI (WAALWIJK and FLAVELL 1978), or high-pressure liquid chromatography (EICK et al. 1983). The latter method in particular is highly sensitive. Adenovirus type 2 (Ad2) virion DNA yielded

<0.04% of 5mC, if any, among its cytidine residues, (Eick et al. 1983). The amount of 6mA was found to be <0.01% among the adenine residues of Ad2 DNA (*Günthert* et al. 1976). Trace amounts of 5mC were occasionally found in Ad2 DNA preparations extracted from virions that had not been treated with DNase beforehand. It is likely that these DNA preparations contained trace amounts of cellular DNA from KB host cells in which Ad2 DNA had been propagated. Thus, the very low levels of 5mC in these viral DNA preparations (Eick et al. 1983) could probably be explained by contamination with tiny amounts of cellular DNA.

Human cell DNA contains 3.6% (KB DNA) to 4.4% (HEK DNA) of 5mC (Günthert et al. 1976). Ad2 DNA grown on these host cells seems to be devoid of 5mC or 6mA, as determined by the most sensitive techniques available. Thus, the interesting problem poses itself of how viral DNA escapes methylation, while the DNA of the host is extensively methylated. Several possible explanations can be considered.

It is interesting to note that human KB cellular DNA inserted into virion DNA and encapsidated into complete Ad12 particles does not appear to be significantly methylated, while the same cellular DNA sequences in human KB cells are strongly methylated (Deuring et al. 1981). The SYREC symmetric recombinants of Ad12 DNA are full-length molecules, identical in size to viral DNA, and constitute recombinants of the left terminus of Ad12 DNA and of human cellular KB DNA (Deuring et al. 1981; Deuring and Doerfler 1983). The example of the recombinant SYREC molecules provides direct evidence for the notion that cellular DNA sequences escape DNA methylation when they are inserted into free viral or virus-like genomes which replicate independently of cellular DNA. Thus, transfer of methyl groups to DNA may depend on the location in a certain nuclear compartment, rather than on the origin of the sequence to be modified.

There are several additional examples of virion DNA not being methylated to detectable levels (Kaye and Winocour 1967; Tjia et al. 1979). In contrast, the DNA of the herpes-like frog virus FV3 is extensively methylated; >20% of all the cytidine residues in FV3 DNA are 5mC (Willis and Granoff 1980).

Experiments using the isoschizomeric restriction enzyme pair *Hpa*II and *Msp*I (Waalwijk and Flavell 1978) demonstrated that the free intracellular viral DNA sequences in productively or abortively infected cells are not extensively methylated either (Vardimon et al. 1980), as no difference was observed in the cleavage patterns of viral DNA by the two enzymes. Ad2 DNA replicates to large copy numbers in human cells, whereas Ad12 DNA cannot replicate in hamster (BHK21) cells (Doerfler 1968, 1969; Doerfler and Lundholm 1970; Fanning and Doerfler 1976; Vardimon et al. 1980). There is no evidence in either system that free parental viral DNA or newly synthesized free viral DNA in productively infected cells becomes methylated at 5'-CCGG-3' sites. Levels of methylation at other 5'-CpG-3'-containing sites in free intracellular viral DNA have not yet been investigated. Similarly, experiments are needed to exclude the possibility that extremely low levels of 5mC could exist even in free viral DNA at 5'-CCGG-3' sites.

We have shown previously that in BHK21 cells abortively infected with Ad12 early viral functions, but not late viral genes, are expressed (ORTIN et al. 1976; ESCHE et al. 1979). Moreover, the regulation of early-versus-late gene expression of adenovirus DNA in productively infected cells functions meticulously, without apparent recourse to DNA methylation as a possible regulatory signal. We cannot rule out the presence of a few methyl groups at decisive regulatory sites in free viral DNA in productively or abortively infected cells, although it is unlikely (U. WIENHUES and W. DOERFLER, unpublished). But viral DNA methylation at a level found in integrated viral genomes in transformed cells (SUTTER et al. 1978; SUTTER and DOERFLER 1979, 1980; KRUCZEK and DOERFLER 1982) simply does not exist in free adenovirus DNA. Thus, it is likely that regulatory principles independent of DNA methylation are operative in free viral DNA. Teleologically speaking, it could be considered inopportune for adenovirus DNA to succumb to long-term regulatory signals of cellular provenience. As can be deduced from work on integrated adenovirus DNA sequences, methylation of integrated viral sequences leads to the (irreversible?) silencing of viral genes. Such a mechanism would obviously be impractical for the regulation of viral genes, and adenovirus DNA must have developed different ways for the temporal regulation of its expression schedule. Of course, the questions remain of how adenovirus DNA regulates expression and how free viral DNA manages to escape DNA methylation.

4 Inverse Correlations Between the Extent of DNA Methylation of Integrated Adenoviral DNA and the Level of Expression of Adenoviral Genes in Transformed Cells

In the course of intensive investigations on the mode of integration of adenovirus DNA in transformed and tumor cells (DOERFLER 1982; DOERFLER et al. 1983 for recent reviews), we have discovered that, in contrast to free intracellular viral or virion DNA, integrated Ad12 DNA in hamster cells is strongly methylated at 5'-CCGG-3' (HpaII) and 5'-CCCGGG-3' (SmaI) sites (SUTTER et al. 1978). Prior to that finding, evidence was adduced that 5'-GTCGAC-3' (SalI) sites were more strongly methylated in DNA from the two Ad12-transformed hamster cell lines T637 and HA12/7 than in DNA from BHK21 cells, the parent line of T637 cells (GRONEBERG et al. 1977). It had also been demonstrated (GÜNTHERT et al. 1976), and was recently confirmed (EICK et al. 1983), that the total 5mC content of DNA from Ad12-transformed T637 and HA12/7 cells (3.7% 5mC) was higher than that of DNA from BHK21 cells (2.5% 5mC). Furthermore, it was shown that both the 5'-CCGG-3' (MspI) sites and the 5'-TCGA-3' (TaqI) sites were more markedly methylated in the DNA from T637 cells than in BHK21 cells (EICK et al. 1983). Thus, evidence is accumulating from different types of analyses that a number of different 5'-CpG-3'-containing sequences are more intensely modified in Ad12-transformed cell lines than in non-virus-transformed cells. Interestingly, however, the levels of DNA methylation in Ad12-induced tumors are strikingly low, as will be discussed later, and these levels increase with continuing passage of these tumor cells in culture (KUHLMANN and DOERFLER 1982).

In addition, the finding that integrated viral DNA sequences are extensively methylated, while virion DNA or free viral DNA in infected cells is not detectably methylated at all, still provides one of the best lines of evidence for the occurrence of de novo methylation of DNA in mammalian cell systems. With the advent of restriction endonuclease pairs diagnostic for DNA methylation at certain nucleotides in specific nucleotide sequences (WAALWIJK and FLAVELL 1978), it became possible to demonstrate methylation of adenovirus DNA sequences in transformed cells directly (SUTTER and DOERFLER 1979, 1980). Since we had previously determined the patterns of expression of the early adenoviral genes in some of the Ad12-transformed hamster lines (ORTIN et al. 1976) and in Ad12-induced rat brain tumor lines (IBELGAUFTS et al. 1980), we were in a good position to investigate relationships between adenovirus DNA methylation and adenoviral gene expression. An inverse correlation between adenoviral gene expression and DNA methylation at 5'-CCGG-3' sites was discovered (SUTTER and DOERFLER 1979, 1980). Similar results were subsequently obtained in Ad2-transformed hamster cells (VARDIMON et al. 1980). Early adenovirus genes were undermethylated in cell lines in which they were expressed; late viral genes usually not expressed in adenovirus-transformed hamster cells were heavily methylated at 5'-CCGG-3' sites. Conversely, early viral genes not expressed in transformed cell lines, such as the E3 region in the Ad12-transformed hamster cell line HA12/7 (ORTIN et al. 1976; SCHIRM and DOERFLER 1981) or the E2a region in the Ad2-transformed hamster cell lines HE2 or HE3 (JOHANSSON et al. 1978; ESCHE 1982), are methylated at some or all of the MspI sites (SUTTER and DOERFLER 1979, 1980; VARDIMON et al. 1980; KRUCZEK and DOERFLER 1982). In some of the Ad12-induced rat brain tumor lines some of the late viral genes are expressed (IBELGAUFTS et al. 1980), and these viral genes are undermethylated at MspI sites (SUTTER and DOERFLER 1980). Comparable inverse correlations frequently, but not invariably, involving the 5'-CCGG-3' sites were subsequently described in many viral and nonviral eukaryotic systems (for reviews see RAZIN and RIGGS 1980; DOERFLER 1981, 1983; EHRLICH and WANG 1981).

Although the Ad12 sequences integrated in cellular DNA in virus-induced tumors were hardly methylated at all at 5'-CCGG-3' or at 5'-GCGC-3' sites, viral DNA sequences did not appear to be expressed in these tumors (KUHLMANN and DOERFLER 1982). We concluded that the absence of DNA methylation probably constituted a necessary, but not a sufficient, precondition for viral gene expression, as had been demonstrated in other systems as well (VAN DER PLOEG and FLAVELL 1980; OTT et al. 1982).

In the same series of investigations we discovered that upon explantation of tumor cells into tissue culture, and upon serial passage, the levels of

DNA methylation in integrated viral genomes at the 5′-CCGG-3′ and 5′-GCGC-3′ sites increased. This shift in the patterns of methylation was gradual, and did not noticeably alter the absence of expression of the persisting viral genomes. The shift was not random, but appeared to follow an ordered schedule, with certain *Msp*I sites becoming methylated earlier than others (KUHLMANN and DOERFLER 1983). We do not know at present what factors affect or regulate the extent of viral DNA methylation in tumors or in transformed cells.

The T637 hamster cells are an Ad12-transformed cell line containing about 22 copies of Ad12 DNA per diploid genome (SUTTER et al. 1978; STABEL et al. 1980). Morphological revertants of this cell line have been isolated and characterized (GRONEBERG et al. 1978; GRONEBERG and DOERFLER 1979; EICK et al. 1980). Some of these revertants contain only one remaining copy of viral DNA or fragments thereof (EICK et al. 1980). In comparison with the levels in cell line T637, viral DNA methylation is markedly increased in the revertants, and the extent of expression is strikingly diminished (EICK et al. 1980; SCHIRM and DOERFLER 1981).

5 Methylation at the Promoter and 5′ Terminus of a Gene Affects Gene Expression

The inverse correlation established in many different eukaryotic systems between the extents of DNA methylation at 5′-CCGG-3′ sites and gene expression does not pertain to DNA methylation along the entire stretch of a gene or a group of genes. Evidence has been adduced from studies on Ad12-transformed hamster cells that 5mC positioned at the 5′-CCGG-3′ sites at the 5′ termini of viral genes, and in the promoter/leader sequences, correlates to the shutoff of integrated viral genes (KRUCZEK and DOERFLER 1982). The data summarized in Fig. 1 demonstrate that some of the 5′-CCGG-3′ (*Msp*I) sites at the 3′ termini in the E1 or the E2a regions of integrated Ad12 genomes in cell lines T637, HA12/7, and A2497-3 are methylated, although these regions are expressed in all three cell lines (ORTIN et al. 1976; SCHIRM and DOERFLER 1981; ESCHE and SIEGMANN 1982). In contrast, in cell line HA12/7 the *Msp*I sites at the 5′ terminus of the E3 region are all methylated (KRUCZEK and DOERFLER 1982); consequently this region fails to be expressed (ORTIN et al. 1976; SCHIRM and DOERFLER 1981; KRUCZEK and DOERFLER 1982). Similar correlations have been observed for the 5′-GCGC-3′ (*Hha*I) sites. Evidence derived from several eukaryotic genes, the adenine phosphoribosyl transferase and dihydrofolate reductase genes (STEIN et al. 1983), the globin gene (FLAVELL et al. 1982), the rat albumin gene (OTT et al. 1982), and the rDNA in *Xenopus laevis* (LA VOLPE et al. 1982), also point to methylation at the 5′ termini of these genes as the decisive shutoff signal. It is unknown at present in what way DNA methylation at highly specific sites blocks the initiation of transcription.

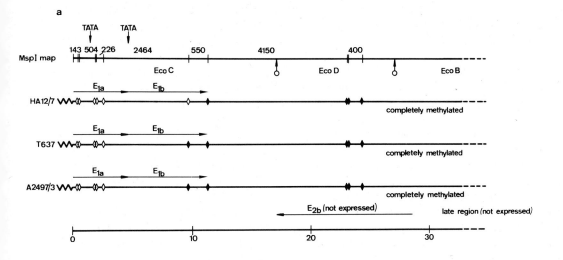

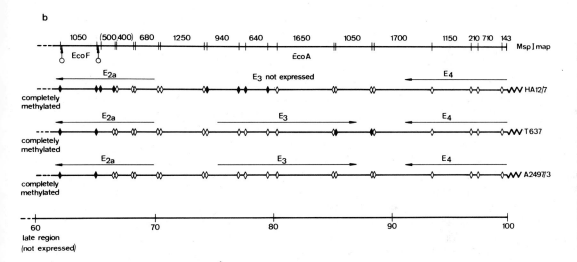

Fig. 1a, b. Functional maps of **a** the left and **b** the right termini of the Ad12 DNA molecules integrated into the DNA of cell lines HA12/7, T637, and A2497-3. The *horizontal lines* represent the Ad12 genomes (——) integrated into the genomes of cell lines (vvv) as indicated. The *Msp*I maps of the left (**a**) and right (**b**) ends of Ad12 DNA are presented in the *top line*. *Vertical bars on* and *figures above* the *horizontal lines* indicate the locations of the *Msp*I sites and the sizes of some of the *Msp*I fragments, respectively. The *arrows* (⟁) designate *Eco*RI sites. *TATA* marks the locations of presumptive Goldberg-Hogness signals in the E1a region. The unmethylated (◇) and methylated (◆) 5′-CCGG-3′ sites in the integrated Ad12 genomes in the lines HA12/7, T637, and A2497-3 are designated by *open* or *closed symbols* as indicated. The *horizontal arrows* indicate map positions and directions of transcription of the individual early regions in each of the cell lines. Presence of *an arrow* indicates expression of corresponding regions of the Ad 12 genome. *Msp*I sites to the right of the *Eco*RI-D/B and to the left of the *Eco*RI-E/F junctions have not been mapped. The early regions of Ad12 DNA and a *fractional length scale* have also been indicated. (From KRUCZEK and DOERFLER 1982)

6 The E2a Region of the Adenovirus Type 2 Genome Is Not Expressed in Some Transformed Cell Lines Although the Entire Gene Persists in These Cell Lines

The Ad2-transformed hamster cell lines HE1, HE2, and HE3 (JOHANSSON et al. 1978; COOK and LEWIS 1979) have served a very useful role in the study of the function of 5mC at specific sites. It has been demonstrated, both by direct immunofluorescence on SDS-polyacrylamide gels with cell extracts (JOHANSSON et al. 1978) and by in vitro translation of hybrid-selected messenger RNAs (ESCHE 1982), that cell lines HE2 and HE3 do not express the Ad2-specific DNA-binding protein, whereas cell line HE1 does express this 72K protein. The DNA-binding protein of Ad2 (VAN DER VLIET and LEVINE 1973) is encoded in the E2a region of the viral genome. All three hamster cell lines contain the intact E2a region within the integrated Ad2 genomes (VARDIMON and DOERFLER 1981). Cell lines HE1, HE2, and HE3 contain 2–4, 2–4, and 6–10 genome fragment equivalents per cell, respectively. We have been able to demonstrate that, in addition to the intact E2a regions, each cell line contains the complete late promoter of the E2a region (VARDIMON et al. 1982b), and that in cell line HE1, in which the DNA-binding protein is synthesized, this late promoter is utilized (VARDIMON et al. 1983). Remarkably, the E2a region is controlled by a complex array of promoter sequences, three promoters being used predominantly early and one region being used predominantly late in a productive infection cycle (CHOW et al. 1979; BAKER and ZIFF 1981). Thus, the failure of the DNA-binding protein to be expressed in cell lines HE2 and HE3 cannot be due simply to a defective gene or to deletions in the promoter region of the E2a segment. In cell lines HE2 and HE3 all 5′-CCGG-3′ sites are methylated; in cell line HE1 the same sites are all unmethylated, as determined by restriction enzyme analysis using HpaII and MspI endonucleases (VARDIMON et al. 1980). Thus, with the E2a region of Ad2 DNA in three transformed hamster cell lines, we have discovered an example of a perfect inverse correlation between DNA methylation at 5′-CCGG-3′ sites and the expression of the DNA-binding protein. It is worth noting that the 5′-GGCC-3′ (HaeIII) sites in the E2a region are not methylated at all (VARDIMON et al. 1982b). It appears that the HaeIII sites do not have any effect on gene expression via DNA methylation.

Findings on the correlation between DNA methylation at specific sites and shutoff of these genes could indicate a causal relationship between DNA methylation and gene expression, or that DNA methylation is a consequence of the absence of gene expression. In order to determine more precisely and in a more direct way the role that DNA methylation can play in the control of gene expression, we have developed systems in which the effects on gene expression of in vitro methylation at specific sites of certain viral segments can be directly investigated. Some of the results are described in the following section. More recently performed experiments will be mentioned briefly at the end of the review.

7 In Vitro Methylation of Cloned Adenoviral Genes and Its Effect on Gene Expression

At this stage of our investigations it could not be determined whether DNA methylation represented the cause or the consequence of gene inactivation. In order to distinguish between these alternatives the following set of experiments was designed. The cloned E2a region of Ad2 DNA was methylated in vitro, using the *Hpa*II DNA methyltransferase (from *Hemophilus parainfluenzae;* MANN and SMITH 1977) which modified the 5'-CCGG-3' sequences at the internal C residues. The application of this prokaryotic DNA-methyltransferase appeared sensible, since in the Ad2-transformed hamster cell lines HE2 and HE3, the E2a region in the integrated Ad2 genomes was completely methylated at the 5'-CCGG-3' sites and the E2a region was not expressed in these cell lines. Conversely, all 5'-CCGG-3' sites of the E2a region were unmodified in cell line HE1, in which the E2a region was expressed. When the in vitro methylated 5'-CmCGG-3' E2a genes of Ad2 DNA were microinjected into nuclei of *Xenopus laevis* (TELFORD et al. 1979; GROSSCHEDL and BIRNSTIEL 1980) the microinjected DNA remained methylated for at least 24 h, and its transcription into mRNA was completely blocked (VARDIMON et al. 1981, 1982a; LANGNER et al. 1984).

On the other hand, unmethylated DNA that was microinjected independently into nuclei of *Xenopus laevis* oocytes was readily expressed as Ad2-specific RNA (VARDIMON et al. 1981, 1982a). It could be shown that the late E2a promoter of unmethylated DNA was utilized in *Xenopus* oocytes. Viral RNA was not detectably initiated on the methylated E2a fragment. Unmethylated sea urchin histone genes coinjected with methylated Ad2 genes continued to be expressed in *Xenopus* oocytes (VARDIMON et al. 1982a). Thus, the transcriptional block of the E2a region of Ad2 DNA was not artifactual, but was somehow caused by methylation at the 5'-CCGG-3' sites (VARDIMON et al. 1981, 1982a). We also methylated the 5'-GGCC-3' sites in the E2a region of Ad2 DNA, i.e., the inverse sequences, by using the *Bsu*RI DNA methyltransferase (from *Bacillus subtilis;* GÜNTHERT et al. 1981). After microinjection into nuclei of *Xenopus laevis,* the *Bsu*RI-methylated E2a region continued to be expressed (VARDIMON et al. 1982b). In cell lines HE1, HE2, and HE3 the 5'-GGCC-3' sites in the E2a regions of integrated Ad2 DNA were not methylated (VARDIMON et al. 1982b). We thus concluded that DNA methylation had to occur at highly specific sequences in order to be functionally relevant. Furthermore, DNA methylation at specific sites (5'-CCGG-3') in some way caused gene inactivation. Similar conclusions were reached by workers investigating other experimental systems (WAECHTER and BASERGA 1982; STEIN et al. 1982; FRADIN et al. 1982). Peculiarly, in the case of the thymidine kinase gene from herpes simplex virus, methylation of adenine at the *Eco*RI site led to gene inactivation upon microinjection into hamster cells (WAECHTER and BASERGA 1982). It cannot be decided at the present time whether the functionally relevant sites for DNA methylation will be the same ones for all genes. It is conceivable that some genes do not respond to methylation at 5′-CCGG-3′ sites, but are regulated by methylation elsewhere. Moreover, it will have to be determined which parts of a gene or of its regulatory sequences have to be methylated for gene inactivation to ensue.

In preliminary experiments we have not observed any difference in the reactivity of antibodies to Z DNA between the cloned HpaII-methylated and the cloned unmethylated E2a region of Ad2 DNA. Thus, there is no evidence at present that upon methylation the E2a DNA would convert to the Z configuration (L. Vardimon and W. Doerfler, unpublished experiments). Of course, such a possibility cannot be ruled out either. As pointed out earlier, DNA methylation might affect gene regulation by modulating the binding of proteins to specific DNA sequences. By no means do I intend to imply that DNA methylation would be the only way to inactivate specific genes.

8   Clusters of 5′-CpG-3′ Sequences

At present, we lack information about the significance of clusters of 5′-CpG-3′ sequences in DNA, in particular of 5′-CCGG-3′ or 5′-GCGC-3′ sequences. It has been noted that clusters of these latter two sequences are distributed nonrandomly. In several instances 5′-CpG-3′ dinucleotides accumulate in front of the regulatory (promoter) sequences of genes (FELSENFELD et al. 1982; DOERFLER et al. 1982). At the left terminus of the DNA of Ad12 such 5′-CpG-3′ clusters and interdigitating 5′-CCGG-3′ and 5′-GCGC-3′ sequences are particularly apparent (DOERFLER 1981). These sequences might constitute signals in their own right; they could be modulated by specific DNA methylations at these sites. There is a striking sequence arrangement –

5′-GGGGCGCGGCGTGGGAG*CCGGGCGC*GCCGGATGTGACG-3′
   126   156

(SUGISAKI et al. 1980; BOS et al. 1981) – in front of the first TATTA sequence of the E1a region in Ad12 DNA. Expression of the E1a region is prerequisite for the transcription of most other early regions of the virion (BERK et al. 1979; JONES and SHENK 1979), hence the promoter region of the E1 segment presumably assumes a pivotal regulatory role. The sequence shown also harbors part of one of the enhancers of the Ela region of Ad12 DNA, CGGATGTGACG (HEARING and SHENK 1983).

9   Evaluation of In Vitro DNA Methylation Experiments

The results of in vitro DNA methylation experiments, using a number of different cloned eukaryotic genes and oocytes from Xenopus laevis or mammalian cells as recipients, demonstrate that 5mC groups at highly specific locations, in sequences which are frequently 5′-CCGG-3′ but may vary from gene to gene, inhibit the transcription of these genes. The mechanism by which gene expression is blocked by the presence of methylated nucleotides is still unknown. As a note of caution it should be conceded that in some of the microinjection experiments using methylated DNA, heterologous cell systems were employed to assay for the expression of unmethylated or methylated genes. Obviously, in heterologous systems control mechanisms may conceivably function in anomalous ways. This reservation has been partly obliterated by showing that the methylated E2a region of Ad2 DNA is not expressed either after transfection into HeLa cells (LANGNER et al. 1984). Moreover, in most of the experiments, the cloned DNA fragment was injected as an insert in a bacterial plasmid; likewise methylation was not confined to the insert or to specific parts of the insert. There is no direct evidence at present that any one of these experimental shortcomings would have affected the outcome of the experiments described in Sect. 7, or would necessitate reinterpretations of the results; however, the experimental design could still be improved upon in future work. This improvement has been achieved in recent work in which only parts of the E2a insert and not the plasmid DNA has been methylated (LANGNER et al. 1984). It would also be desirable to develop an in vitro transcription system for cloned DNA fragments, in which gene expression becomes sensitive to DNA methylation. Previously negative attempts in this respect might be accounted for by failures of the currently employed in vitro transcription systems to contain all the factors required in regulation.

10 DNA-Methyltransferase Activities in Extracts of Adenovirus-Infected and Uninfected Human Cells

The DNA isolated from highly purified adenovirions does not seem to contain detectable amounts of 5mC or 6mA (GÜNTHERT et al. 1976; SUTTER et al. 1978; SUTTER and DOERFLER 1979, 1980; VON ACKEN et al. 1979; DOERFLER et al. 1982; EICK et al. 1983). In contrast, viral DNA integrated into cellular DNA is methylated at specific sites in highly specific patterns (SUTTER and DOERFLER 1979, 1980; VARDIMON et al. 1980; KUHLMANN and DOERFLER 1982; KRUCZEK and DOERFLER 1982). The question of how adenovirus DNA replicating in the nucleus of infected cells can escape being methylated is still unresolved. Several possibilities have to be considered:

1. The activity of DNA methyltransferase(s) in the host cell could be altered after infection.
2. Host DNA methyltransferase(s) might be sequestered into cellular chromatin and not have access to free viral DNA replicating in the nucleus. It is also important to recall that free viral DNA replicates at a very high rate in permissive cell systems.
3. The methyl donor in the methyl transfer reaction, S-adenosylmethionine (AdoMet), may be available in limited supply in adenovirus-infected cells.

   More complex explanations are also possible.
   We have set out to investigate some of these possibilities and have compared the DNA-methyltransferase activities in uninfected and Ad2-infected

HeLa cells. In nuclear extracts of these cell systems, no difference in DNA-methyltransferase activities could so far be observed (B. KNUST-KRON; D. EICK, and W. DOERFLER, unpublished experiments; DOERFLER et al. 1982). It could be clearly demonstrated that these nuclear extracts were capable of methylating adenovirus DNA or salmon sperm DNA de novo. Double-stranded DNA was a substrate, single-stranded DNA was not. Using [^{3}H]-methyl-labeled SAM as methyl donor, [^{3}H]-methyl was transferred to adenovirus DNA. We are currently investigating whether the cell extracts contain different types of DNA-methyltransferases. The possibility exists that the specificity of a single DNA-methyltransferase in KB cells is controlled by host factors which are lost upon extraction of the methyltransferase from the nuclei.

11 Work in Progress and Main Conclusions

The results presented in this review derive mainly from work with human adenoviruses and adenovirus-transformed cells. A very large volume of evidence from many different eukaryotic and also prokaryotic systems (for a current review, see DOERFLER 1983) supports the conclusion that DNA methylations at highly specific sites at the 5' end of genes and/or in the promoter regions of genes directly affect gene expression. The biochemical mechanism by which this regulatory effect is exerted is presently unknown. Strategically positioned methylated bases may influence the binding of specific proteins at certain important sites of DNA directly or via structural alterations of DNA (BEHE and FELSENFELD 1981). Some of the major problems perhaps amenable to experimental analysis in current research can be summarized as follows:

1. The patterns of methylation at all 5'-CpG-3' sites of a gene and its regulatory sequences need to be determined.
2. The functional significance, if any, of each methylated base in DNA should be elucidated.
3. Studies on the regulation of DNA methylation are needed. If DNA methylation is indeed decisive in the long-term regulation of gene activity, a very important role must be ascribed to DNA-methyltransferases (maintenance or de novo) and factors affecting their activity.
4. Can DNA be actively demethylated by demethylating enzymes or proteins?
5. Which 5mC residues are the functionally significant ones?
6. What are the rules in positioning 5mC residues relative to other regulatory signals?
7. What structural changes of DNA can 5mC elicit and/or stabilize, and are structural changes of DNA decisive as regulatory signals?
8. How does 5mC in DNA affect DNA-protein interactions?
9. What functions other than transcription can be affected by 5mC?

A method is urgently needed for determining all 5mC residues within or in the vicinity of a gene and for establishing their functions.

Finally, there are genes that do not appear to subscribe to regulation by DNA methylation, e.g., genes in most free viral genomes. Studies on the methylation pattern of the α2 collagen gene in DNA from five cell types of the chicken reveal that expression of the α2 collagen gene seems to be independent of methylation at 5′-CCGG-3′ sites (McKEON et al. 1982); similar findings were reported for the vitellogenin gene in *Xenopus* (GERBER-HUBER et al. 1983) which purportedly is fully methylated and yet expressed. Of course, it is essential in these analyses to investigate the promoter sites and the correct 5′-CpG-3′-containing sequences, before conclusions can be drawn. Thus, DNA methylation is an element in long-term gene inactivation; it may not be used invariably as a regulatory signal, and it may occasionally assume other functions as well.

We continue to investigate the adenovirus system and are presently pursuing experiments to:

1. Characterize and purify DNA-methyltransferase activities from uninfected and adenovirus-infected human cells, as well as from adenovirus-transformed cells (B. KNUST-KRON and W. DOERFLER, unpublished results).
2. Link the promoter region of the E1a or protein IX genes of Ad12 to the chloramphenicol-acetyltransferase gene in the pSVO-CAT vector (GORMAN et al. 1982) and study the effect of specific methylations in the E1a or protein IX promoter regions (KRUCZEK and DOERFLER, 1983). The results of these investigations have revealed that critical site promoter methylations lead to inactivation of the promoter. The methylation of two 5′-CCGG-3′ sites or of 5′-GCGC-3′ sites upstream of the TATA signal in the E1a promoter of Ad12 DNA inactivated this promoter. Methylation of one 5′-CCGG-3′ or one 5′-GCGC-3′ site downstream of the TATA signal in the promoter of the protein IX gene did not inactivate this promoter.
3. Methylate either the 5′ end or the 3′ body of the cloned E2a region of Ad2 DNA and investigate the effect of methylation at either site on the expression of the E2a region upon microinjection into *Xenopus laevis* oocytes (LANGNER et al. 1984). The methylation of three 5′-CCGG-3′ sites, one 215 base pairs (bp) upstream from the E2a cap site of Ad2 DNA and one each 5 and 23 bp downstream from this site, transcriptionally inactivated the E2a region after microinjection into *Xenopus laevis* oocytes. Methylation of 11 5′-CCGG-3′ sites in the main part of the E2a gene affected its transcriptional activity only slightly, if at all.

Acknowledgments. I am indebted to GERTRUD DEUTSCHLÄNDER for typing this manuscript. Research in the author's laboratory was supported by the *Deutsche Forschungsgemeinschaft* through SFB 74-C1, and by the Ministry of Science and Research of Nordrhein-Westfalen (IVB5-FA9227).

References

Baker CC, Ziff EB (1981) Promoters and heterogeneous 5′ termini of the messenger RNAs of adenovirus serotype 2. J Mol Biol 149:189−221

Behe M, Felsenfeld G (1981) Effects of methylation on a synthetic polynucleotide: the B-Z transition in poly(dG-m^{5}dC)·poly(dG-m^{5}dC). Proc Natl Acad Sci USA 78:1619–1623

Berk AJ, Lee F, Harrison T, Williams J, Sharp PA (1979) Pre-early adenovirus 5 gene product regulates synthesis of early viral messenger RNAs. Cell 17:935–944

Bird AP (1978) Use of restriction enzymes to study eukaryotic DNA methylation pattern. II. The symmetry of methylated sites supports semi-conservative copying of the methylation pattern. J Mol Biol 118:49–60

Bos JL, Polder LJ, Bernards R, Schrier PI, van den Elsen PJ, van der Eb AJ, van Ormondt H (1981) The 2.2 kb E1b mRNA of human Ad12 and Ad5 codes for two tumor antigens starting at different AUG triplets. Cell 27:121–131

Burdon RH, Adams RLP (1980) Eukaryotic DNA methylation. Trends Biochem Sci 5:294–297

Busslinger M, Hurst J, Flavell RA (1983) DNA methylation and the regulation of globin gene expression. Cell 34:197–206

Chow LT, Broker TR, Lewis JB (1979) Complex splicing patterns of RNAs from the early regions of adenovirus-2. J Mol Biol 134:265–303

Christman JK, Price P, Pedrinan L, Acs G (1977) Correlation between hypomethylation of DNA and expression of globin genes in Friend erythroleukemia cells. Eur J Biochem 81:53–61

Christman JK, Weich N, Schoenbrun B, Schneiderman N, Acs G (1980) Hypomethylation of DNA during differentiation of Friend erythroleukemia cells. J Cell Biol 86:366–370

Cohen JC (1980) Methylation of milk-borne and genetically transmitted mouse mammary tumor virus proviral DNA. Cell 19:653–662

Cold Spring Harbor Symposium on Quantitative Biology (1982) vol 47. Cold Spring Harbor, New York

Conner BN, Takano T, Tanaka S, Itakura K, Dickerson RE (1982) The molecular structure of d(^{I}CpCpGpG), a fragment of right-handed double helical A-DNA. Nature 295:294–299

Constantinides PG, Jones PA, Gevers W (1977) Functional striated muscle cells from non-myoblast precursors following 5-azacytidine treatment. Nature (Lond) 267:364–366

Cook JL, Lewis AM Jr (1979) Host response to adenovirus 2-transformed hamster embryo cells. Cancer Res 39:1455–1461

Cooper DN (1983) Eukaryotic DNA methylation. Hum Genet 64:315–333

Creusot F, Acs G, Christman JK (1982) Inhibition of DNA methyltransferase and induction of Friend erythroleukemia cell differentiation by 5-azacytidine and 5-aza-2′-deoxycytidine. J Biol Chem 257:2041–2048

Desrosiers RC, Mulder C, Fleckenstein B (1979) Methylation of herpesvirus saimiri DNA in lymphoid tumor cell lines. Proc Natl Acad Sci USA 76:3839–3843

Deuring R, Klotz G, Doerfler W (1981) An unusual symmetric recombinant between adenovirus type 12 DNA and human cell DNA. Proc Natl Acad Sci USA 78:3142–3146

Deuring R, Doerfler W (1983) Proof of recombination between viral and cellular genomes in human KB cells productively infected by adenovirus type 12: structure of the junction site in a symmetric recombinant (SYREC). Gene 26:283–289

Dickerson RE, Drew HR, Conner BN, Wing RM, Fratini AV, Kopka ML (1982) The anatomy of A-, B-, and Z-DNA. Science 216:475–485

Doerfler W (1968) The fate of the DNA of adenovirus type 12 in baby hamster kidney cells. Proc Natl Acad Sci USA 60:636–643

Doerfler W (1969) Nonproductive infection of baby hamster kidney cells (BHK21) with adenovirus type 12. Virology 38:587–606

Doerfler W (1981) DNA methylation – A regulatory signal in eukaryotic gene expression. J Gen Virol 57:1–20

Doerfler W (1982) Uptake, fixation and expression of foreign DNA in mammalian cells: the organization of integrated adenovirus DNA sequences. Curr Top Microbiol Immunol 101:127–194, Springer Verlag, Berlin Heidelberg New York

Doerfler W (1983) DNA methylation and gene activity. Ann Rev Biochem 52:93–124

Doerfler W, Lundholm U (1970) Absence of replication of the DNA of adenovirus type 12 in BHK21 cells. Virology 40:754–757

Doerfler W, Kruczek I, Eick D, Vardimon L, Kron B (1982) DNA methylation and gene activity: the adenovirus system as a model. Cold Spring Harbor Symp Quant Biol 47:593–603

Doerfler W, Gahlmann R, Stabel S, Deuring R, Lichtenberg U, Schulz M, Eick D, Leisten R (1983) On the mechanism of recombination between adenoviral and cellular DNAs: The structure of junction sites. Curr Top Microbiol Immunol 109:193–228

Ehrlich M, Wang RY-H (1981) 5-methylcytosine in eukaryotic DNA. Science 212:1350–1357

Eick D, Stabel S, Doerfler W (1980) Revertants of adenovirus type 12-transformed hamster cell line T637 as tools in the analysis of integration patterns. J Virol 36:41–49

Eick D, Fritz H-J, Doerfler W (1983) Quantitative determination of 5-methylcytosine in DNA by reverse-phase high-performance liquid chromatography. Anal. Biochem. 135:165–167

Esche H (1982) Viral gene products in adenovirus type 2-transformed hamster cells. J Virol 41:1076–1082

Esche H, Siegmann B (1982) Expression of early viral gene products in adenovirus type 12-infected and -transformed cells. J Gen Virol 60:99–113

Esche H, Schilling R, Doerfler W (1979) In vitro translation of adenovirus type 12-specific mRNA isolated from infected and transformed cells. J Virol 30:21–31

Fanning E, Doerfler W (1976) Intracellular forms of adenovirus DNA. V. Viral DNA sequences in hamster cells abortively infected and transformed with human adenovirus type 12. J Virol 20:373–383

Felsenfeld G, Nickol J, Behe M, McGhee J, Jackson D (1982) Methylation and chromatin structure. Cold Spring Harbor Symp Quant Biol 47:577–584

Flavell RA, Grosveld F, Busslinger M, de Boer E, Kioussis D, Mellor AL, Golden L, Weiss E, Hurst J, Bud H, Bullman H, Simpson E, James R, Townsend ARM, Taylor PM, Schmidt W, Ferluga J, Leben L, Santamaria M, Atfield G, Festenstein H (1982) Structure and expression of the human globin genes and murine histocompatibility antigen genes. Cold Spring Harbor Symp Quant Biol 47:1067–1078

Fradin A, Manley JL, Prives CL (1982) Methylation of simian virus 40 *Hpa*II site affects late, but not early, viral gene expression. Proc Natl Acad Sci USA 79:5142–5146

Gerber-Huber S, May FEB, Westley BR, Felber BK, Hosbach HA, Andres A-C, Ryffel GU (1983) In contrast to other *Xenopus* genes the estrogen-inducible vitellogenin genes are expressed when totally methylated. Cell 33:43–51

Gorman CM, Moffat LF, Howard BH (1982) Recombinant genomes which express chloramphenicol acetyltransferase in mammalian cells. Mol Cell Biol 2:1044–1051

Groneberg J, Doerfler W (1979) Revertants of adenovirus type 12-transformed hamster cells have lost part of the viral genomes. Int J Cancer 24:67–74

Groneberg J, Chardonnet Y, Doerfler W (1977) Integrated viral sequences in adenovirus type 12-transformed hamster cells. Cell 10:101–111

Groneberg J, Sutter D, Soboll H, Doerfler W (1978) Morphological revertants of adenovirus type 12-transformed hamster cells. J Gen Virol 40:635–645

Grosschedl R, Birnstiel ML (1980) Identification of regulatory sequences in the prelude sequences of an H2A histone gene by the study of specific deletion mutants in vivo. Proc Natl Acad Sci USA 77:1432–1436

Groudine M, Eisenman R, Weintraub H (1981) Chromatin structure of endogenous retroviral genes and activation by an inhibitor of DNA methylation. Nature (Lond) 292:311–317

Guntaka RV, Rao PY, Mitsialis SA, Katz R (1980) Modification of avian sarcoma proviral DNA sequences in nonpermissive XC cells but not in permissive chicken cells. J Virol 34:569–572

Günthert U, Schweiger M, Stupp M, Doerfler W (1976) DNA methylation in adenovirus, adenovirus-transformed cells, and host cells. Proc Natl Acad Sci USA 73:3923–3927

Günthert U, Jentsch S, Freund M (1981) Restriction and modification in *Bacillus subtilis*: Two DNA methyltransferases with *Bsu*RI specificity. II. Catalytic properties, substrate specificity, and mode of action J Biol Chem 256:9346–9351

Harris M (1982) Induction of thymidine kinase in enzyme-deficient Chinese hamster cells. Cell 29:483–492

Hattman S (1981) DNA methylation. In: Boyer PD (ed) The enzymes. Academic, New York, pp 517–547

Hearing P, Shenk T (1983) The adenovirus type 5 EIA transcriptional control region contains a duplicated enhancer element. Cell 33:695–703

Holliday R, Pugh JE (1975) DNA modification mechanisms and gene activity during development. Developmental clocks may depend on the enzymic modification of specific bases in repeated DNA sequences. Science 187:226–232

Ibelgaufts H, Doerfler W, Scheidtmann KH, Wechsler W (1980) Adenovirus type 12-induced rat tumor cells of neuroepithelial origin: persistence and expression of the viral genome. J Virol 33:423–437

Jack RS, Brown MT, Gehring WJ (1982) Protein blotting as a means to detect sequence-specific DNA-binding proteins. Cold Spring Harbor Symp Quant Biol 47:483–491

Johansson K, Persson H, Lewis AM, Pettersson U, Tibbetts C, Philipson L (1978) Viral DNA sequences and gene products in hamster cells transformed by adenovirus type 2. J Virol 27:628–639

Jones N, Shenk T (1979) An adenovirus type 5 early gene function regulates expression of other early viral genes. Proc Natl Acad Sci USA 76:3665–3669

Jones PA, Taylor SM (1981) Hemimethylated duplex DNAs prepared from 5-azacytidine-treated cells. Nucleic Acids Res 9:2933–2947

Jones PA, Taylor SM, Mohandas T, Shapiro LJ (1982) Cell cycle-specific reactivation of an inactive X-chromosome locus by 5-azadeoxycytidine. Proc Natl Acad Sci USA 79:1215–1219

Kaye AM, Winocour E (1967) On the 5-methyl-cytosine found in the DNA extracted from polyomavirus. J Mol Biol 24:475–478

Kruczek I, Doerfler W (1982) The unmethylated state of the promoter/leader and 5′ regions of integrated adenovirus genes correlates with gene expression. The EMBO J 4:409–414

Kruczek I, Doerfler W (1983) Expression of the chloramphenicol acetyltransferase gene in mammalian cells under the control of adenovirus type 12 promoters: Effect of promoter methylation on gene expression. Proc. Natl. Acad. Sci. USA 80:7586–7590

Kuhlmann I, Doerfler W (1982) Shifts in the extent and patterns of DNA methylation upon explantation and subcultivation of adenovirus type 12-induced hamster tumor cells. Virology 118:169–180

Kuhlmann I, Doerfler W (1983) Loss of viral genomes from hamster tumor cells and nonrandom alterations in patterns of methylation of integrated adenovirus type 12 DNA. J Virol 47:631–636

Mandel JL, Chambon P (1979) DNA methylation: organ-specific variations in the methylation pattern within and around ovalbumin and other chicken genes. Nucleic Acids Res 7:2081–2103

Langner K-D, Vardimon L, Renz D, Doerfler W (1984) DNA methylations of three 5′-CCGG-3′ sites in the promoter and 5′ regions inactivate the E2a gene of adenovirus type 2. Proc Natl Acad Sci USA 81

La Volpe A, Taggart M, Macleod D, Bird A (1982) Coupled demethylation of sites in a conserved sequence of *Xenopus* ribosomal DNA. Cold Spring Harbor Symp Quant Biol 47:585–592

Mann MB, Smith HO (1977) Specificity of *Hpa*II and *Hae*III methylases. Nucleic Acids Res 4:4211–4221

McGhee JD, Ginder GD (1979) Specific DNA methylation sites in the vicinity of the chicken β-globin gene. Nature Lond 280:419–420

McGhee JD, Felsenfeld G (1980) Nucleosome structure. Annu Rev Biochem 49:1115–1156

McKeon C, Ohkubo H, Pastan I, de Crombrugghe B (1982) Unusual methylation pattern of the α2(I) collagen gene. Cell 29:203–210

Ortin J, Scheidtmann KH, Greenberg R, Westphal M, Doerfler W (1976) Transcription of the genome of adenovirus type 12. III. Maps of stable RNA from productively infected human cells and abortively infected and transformed hamster cells. J Virol 20:355–372

Ott M-O, Sperling L, Cassio D, Levilliers J, Sala-Trepat J, Weiss MC (1982) Undermethylation at the 5′ end of the albumin gene is necessary but not sufficient for albumin production by rat hepatoma cells in culture. Cell 30:825–833

Razin A, Riggs AD (1980) DNA methylation and gene function. Science 210:604–610

Razin A, Friedman J (1981) DNA methylation and its possible biological roles. Prog Nucleic Acids Res Mol Biol 25:33–52

Riggs AD (1975) X inactivation, differentiation and DNA methylation. Cytogenet Cell Genet 14:9–25

Sager R, Kitchin R (1975) Selective silencing of eukaryotic DNA. Science 189:426–433

Schirm S, Doerfler W (1981) Expression of viral DNA in adenovirus type 12-transformed cells, in tumor cells, and in revertants. J Virol 39:694–702

Smith SS, Yu JC, Chen CW (1982) Different levels of DNA modification at 5'-CCGG in murine erythroleukemia cells and the tissues of normal mouse spleen. Nucleic Acids Res 10:4305–4320

Stabel S, Doerfler W, Friis RR (1980) Integration sites of adenovirus type 12 DNA in transformed hamster cells and hamster tumor cells. J Virol 36:22–40

Stein R, Razin A, Cedar H (1982) In vitro methylation of the hamster adenine phosphoribosyltransferase gene inhibits its expression in mouse L cells. Proc Natl Acad Sci USA 79:3418–3422

Stein R, Sciaki-Gallili N, Razin A, Cedar H (1983) Pattern of methylation of two genes coding for housekeeping functions. Proc Natl Acad Sci USA 80:2422–2426

Sugisaki H, Sugimoto K, Takanami M, Shiroki K, Saito I, Shimojo H, Sawada Y, Uemizu Y, Uesugi S, Fujinaga K (1980) Structure and gene organization in the transforming HindIII-G fragment of Ad12. Cell 20:777–786

Sutter D, Doerfler W (1979) Methylation of integrated viral DNA sequences in hamster cells transformed by adenovirus 12. Cold Spring Harbor Symp Quant Biol 44:565–568

Sutter D, Doerfler W (1980) Methylation of integrated adenovirus type 12 DNA sequences in transformed cells is inversely correlated with viral gene expression. Proc Natl Acad Sci USA 77:253–256

Sutter D, Westphal M, Doerfler W (1978) Patterns of integration of viral DNA sequences in the genomes of adenovirus type 12-transformed hamster cells. Cell 14:569–585

Taylor SM, Jones PA (1979) Multiple new phenotypes induced in 10T1/2 and 3T3 cells treated with 5-azacytidine. Cell 17:771–779

Telford JL, Kressmann A, Koski RA, Grosschedl R, Müller F, Clarkson SG, Birnstiel ML (1979) Delimitation of a promoter for RNA polymerase III by means of a functional test. Proc Natl Acad Sci USA 76:2590–2594

Tjia S, Carstens EB, Doerfler W (1979) Infection of Spodoptera frugiperda cells with Autographa californica nuclear polyhedrosis virus. II. The viral DNA and the kinetics of its replication. Virology 99:399–409

Van der Ploeg LHT, Flavell RA (1980) DNA methylation in the human γδβ-globin locus in erythroid and nonerythroid tissues. Cell 19:947–958

Van der Ploeg LHT, Groffen J, Flavell RA (1980) A novel type of secondary modification of two CCGG residues in human γδβ-globin gene locus. Nucleic Acids Res 8:4563–4574

Van der Vliet PC, Levine AJ (1973) DNA-binding proteins specific for cells infected by adenovirus. Nature (New Biol) 246:170–174

Vardimon L, Doerfler W (1981) Patterns of integration of viral DNA in adenovirus type 2-transformed hamster cells. J Mol Biol 147:227–246

Vardimon L, Neumann R, Kuhlmann I, Sutter D, Doerfler W (1980) DNA methylation and viral gene expression in adenovirus-transformed and -infected cells. Nucleic Acids Res 8:2461–2473

Vardimon L, Kuhlmann I, Cedar H, Doerfler W (1981) Methylation of adenovirus genes in transformed cells and in vitro: influence on the regulation of gene expression. Eur J Cell Biol 25:13–15

Vardimon L, Kressmann A, Cedar H, Maechler M, Doerfler W (1982a) Expression of a cloned adenovirus gene is inhibited by in vitro methylation. Proc Natl Acad Sci USA 79:1073–1077

Vardimon L, Günthert U, Doerfler W (1982b) In vitro methylation of the BsuRI (5'-GGCC-3') sites in the E2a region of adenovirus type 2 DNA does not affect expression in Xenopus laevis oocytes. Mol Cel Biol 2:1574–1580

Vardimon L, Renz D, Doerfler W (1983) Can DNA methylation regulate gene expression? Recent Results in Cancer Research., vol 84. Springer, Heidelberg, pp 90–102

Von Acken U, Simon D, Grunert F, Döring HP, Kröger H (1979) Methylation of viral DNA in vivo and in vitro. Virology 99:152–157

Waalwijk C, Flavell RA (1978) MspI, an isoschizomer of HpaII which cleaves both unmethylated and methylated HpaII sites. Nucleic Acids Res 5:3231–3236

Waechter DE, Baserga R (1982) Effect of methylation on expression of microinjected genes. Proc Natl Acad Sci USA 79:1106–1110

Weintraub H, Larsen A, Groudine M (1981) α-Globin gene switching during the development of chicken embryos: expression and chromosome structure. Cell 24:333–344

Willis DB, Granoff A (1980) Frog virus 3 DNA is heavily methylated at CpG sequences. Virology 107:250–257

Zimmerman SB (1982) The three-dimensional structure of DNA. Annu Rev Biochem 51:395–427

Replicative DNA Methylation in Animals and Higher Plants

B.F. Vanyushin

1 Nature of Minor Bases in DNA

5-Methylcytosine (5mC) has been detected in DNA of all animals and higher plants (Wyatt 1951; Vanyushin and Belozersky 1959; Vanyushin et al. 1970, 1971, 1973a). N^6-methyladenine (6mA) has been shown to exist in DNA of higher plants (Vanyushin et al. 1971; Buryanov et al. 1972), protozoa (Gorovsky et al. 1973; Cummings et al. 1974; Kirnos et al. 1980), fungi (Buryanov et al. 1970), algae (Pakhomova et al. 1968; Rae 1976; Hattman et al. 1978), and invertebrates (Vanyushin et al. 1970), as well as in DNA of prokaryotes (Vanyushin et al. 1968). In DNA of some algae 5-hydroxymethyluracil has been found (Rae 1976). Attempts at finding 6mA in vertebrate DNA and an adenine DNA methylase in nuclei or mitochondria of vertebrates have been unsuccessful (Vanyushin et al. 1970; Kudryashova and Vanyushin 1976). It seems that no essential methylation of the adenine residues in vertebrate DNA takes place. On the other hand, there is no doubt that 6mA detected in DNA of higher plants and other eukaryotes originates from selective methylation of some adenine residues at the polynucleotide level, as is the case in prokaryotes. In contrast to

A.N. Belozersky Laboratory of Molecular Biology and Bioorganic Chemistry, Moscow State University, Moscow 117234, USSR

Current Topics in Microbiology and Immunology, Vol. 108
©Springer-Verlag Berlin · Heidelberg 1984

prokaryotes, adenine methylases in eukaryotes have not been described and the nature of the DNA sites methylated by them is as yet unknown.

2 The Methylated Sequences

In animal (vertebrate) DNA, 5mC occurs mainly in the 5mCG sequence. Comparative studies of various animal DNAs has revealed the following regularities in the intragenomic distribution of 5mC: (a) the genomes of all the animal species studied are methylated unevenly, with 5mC being present in all types of DNA nucleotide sequences that differ in repetition degree; (b) the greatest amount of 5mC is contained in highly repetitive sequences (especially palindromes, satellite DNAs), whereas unique sequences are methylated least of all; (c) different fractions of moderately repetitive sequences may sharply differ in the methylation level (SCHNEIDER-MANN and BILLEN 1973; HARBERS et al. 1975; ROMANOV and VANYUSHIN 1981). These basic features of genome methylation in vertebrates can be explained by intragenomic distribution of CG (5mCG) sequences (ROMANOV and VANYUSHIN 1981). Nevertheless, the methylation of DNA in animals is not confined to the CG sequence. In mouse DNA, 5mC was found in all cytosine-containing dipyrimidine sequences, including 5mC5mC (SALO-MON and KAYE 1970). SNEIDER (1972) reports that 5mCT dinucleotide was isolated from the Novikoff hepatoma cell DNA. Besides, in this DNA 5mC occurs not only in the C5mCGG but also in the 5mCCGG sequence (SNEIDER 1980). In human globin DNA, 5mC seems to occur in the 5mCCGG or 5mC5mCGG sequences (VAN DER PLOEG and FLAVELL 1980). All these facts may be regarded as favoring the existence of a multicomponential system of DNA methylases with different recognition sites in the nucleus of the animal cell. It has been shown that in lymphocyte nuclei isolated from cows suffering from chronic lympholeukosis a new DNA methylase may appear; this enzyme(s) differs from DNA methylase of normal lymphocytes in physicochemical properties and recognition site (BURT-SEVA et al. 1978).

In higher plant DNA, only a fraction of 5mC is localized in 5mCG, an essential amount of this base being found in the middle and at the 5′ ends of pyrimidine oligonucleotides (VANYUSHIN et al. 1962; DOHEIM et al. 1978; KIRNOS et al. 1981). In lily of the valley and wheat DNA, 5mC has been found in the Pu-5mC-Pu, 5mCT, and 5mC5mC sequences, and in a CCC trinucleotide in which two of the cytosine residues are methylated (KIRNOS et al. 1981). A high degree of cytosine residue methylation (>80%) has been established as existing in the CAG and CTG sequences isolated from wheat germ DNA (GRUENBAUM et al. 1981). It was suggested that in plant cells there may be a few different DNA methylases (DOHEIM et al. 1978), which methylate the cytosine residues in the sequences with at least two different types of symmetry: in the first one the rotation axis seems to pass through the CpG dinucleotide ($\overset{*}{\underset{}{\text{CpG}}}_{\text{GpC}}$) and in the second one, through the complementary $\frac{A}{T}$ or $\frac{G}{C}$ base pairs ($\overset{*}{\underset{G}{C}}\,\underset{A}{T}\,\overset{}{\underset{*}{\underset{C}{G}}}$ or $\overset{*}{\underset{G}{C}}\,\underset{G}{C}\,\underset{*}{\underset{C}{G}}$; the asterisk stands for a CH_3 group) (KIRNOS et al. 1981). The same may also be true of DNA methylation in mammalian cells.

3 The 5-Methylcytosine Content

3.1 Species Specificity

The amount of 5mC in total animal DNA varies from 0.1 to 3.0 mol% (WYATT 1951; VANYUSHIN et al. 1970, 1973a). DNA of invertebrates is less methylated than vertebrate DNA. In insects, 5mC seems to be localized in sequences other than CCGG and GCGC. The very high (up to 10 mol%) 5 mC content is a specific feature of most higher plant DNAs (VANYUSHIN and BELOZERSKY 1959; VANYUSHIN et al. 1971; DEUMLING 1981). DNA is less methylated in the archegonial plants than in the flowering plants. The maximal variability of the 5mC content was found in DNA of Liliaceae, one of the ancient plant families. Thus, evolution and species differentiation in plants and animals are reflected in the methylation pattern of their DNA.

3.2 Tissue (Cell) Differences

By direct spectrophotometric determinations of the 5mC content it has been established that DNA from various tissues in an animal and a higher plant differ in the degree of methylation (BERDYSHEV et al. 1967; VANYUSHIN et al. 1970, 1973a; GUSKOVA et al. 1977; CHVOJKA et al. 1978; ROMANOV and VANYUSHIN 1981). The same conclusion was drawn from the measurement of the 5mC radioactivity which animal DNA had acquired after incorporation of labeled pyrimidine precursors (KAPPLER 1971). Tissue differences in the DNA methylation have also been detected by the analysis of the DNA fragments produced by HpaII, MspI, Hha and other CG-specific restriction endonucleases (WAALWIJK and FLAVELL 1978; MCGHEE and GINDER 1979; VAN DER PLOEG and FLAVELL 1980; RAZIN and RIGGS 1980; BIRD et al. 1981; DOERFLER 1981; EHRLICH and WANG 1981).

3.3 Age Differences

The data on the base composition of DNA from somatic cells of spawning Pacific (humpback) salmon seem to be the earliest indication of the existence of age differences in the DNA methylation in animals: during spawning the 5mC content in DNA of various tissues decreases about one and a half- to twofold (BERDYSHEV et al. 1967). In mammals (rat, mouse, cow), a tissue-specific decrease in the DNA methylation level has also been observed (VANYUSHIN et al. 1973b; ZINKOVSKAYA et al. 1978; VANYUSHIN and ROMANENKO 1979). The maximal 5mC content is in DNA of the embryo tissues; it decreases sharply soon after the birth of an animal, then in some organs (liver) it does not change, but in others (brain, heart muscle) it continues to decrease (by 30%–40%) with age. The age-associated decrease in the DNA methylation level is not random: it appears mainly in repeated but not unique DNA sequences (Romanov and Vanyushin 1981) and is predominantly concerned with Pu-5mC-Pu sites (Zinkovskaya et al. 1978). In the cells of old animals, nuclear DNA retains sequences which may potentially be methylated: when methylated in vitro by the homologous DNA methylase, these DNA accept many more CH_3 groups than the corresponding DNA from young and adult animals (Kudryashova and Vanyushin 1976). Under the action of geroprotectors (antioxidants), the DNA methylation level in the cells of aged animals may happen to be similar to that in young ones (Romanenko et al. 1979, 1981). Unfortunately, the mechanisms of an essential DNA demethylation in aged animals are not known; it may be associated with deficiency in S-adenosylmethionine as well as with decrease in methylase activity. In any case, the age-associated changes in DNA methylation may somehow be responsible for the known distortions in DNA replication and transcription in aged cells.

It has been proposed that the methylation degree of DNA should be maximal in zygotes, and during later development DNA is gradually demethylated (Razin and Riggs 1980). Most of the data on the 5mC content in DNA of various animals in ontogenesis do not contradict this proposal. For example, the CpG sites in the spacer of the *Xenopus laevis* rDNA in sperm, and probably in oocytes, are fully methylated; the initial loss of methyl groups from these rDNA sites precedes the gastrula stage and this process continues until the early neurula stage (Bird et al. 1981). All the CCGG sites in the human globin gene regions are fully methylated in sperm DNA but not in somatic tissues (van der Ploeg and Flavell 1980). DNA of chicken sperm contains some very long segments where methylation consecutively affects every CCGG; only some of these sites are free of methylation in somatic cells (Sobieski and Eden 1981). The methylation degree of the CG sites in rabbit trophoblast DNA decreases during ontogenesis (Manes and Menzel 1981). On the other hand, no change has been found in 5mC content in DNA of developing sea urchin in the stages from 64–168 cells onward (Pollock et al. 1978). The histone genes in the sea urchin (*Echinus esculenta*) genome remain unmethylated both in early embryos and in sperm and gastrulae (Bird et al. 1979).

In plants, the 5mC content in total DNA diminishes by about 15% on germination, and in wheat seedlings 5mC begins to appear in sequences other than in seed DNA (Sulimova et al. 1978). This may be due to the new DNA methylases that appear on germination, and is in agreement with our proposal about multiplicity of DNA methylases in plant nucleus (Doheim et al. 1978; Kirnos et al. 1981). This interchange in the methylation of certain DNA sequences may have a regulatory significance and seems to be associated with the known switching over of gene transcription during germination.

4 Methylation and Synthesis of DNA: Cell Cycle

It has been established that methylation of DNA in bacteria takes place during its replication (BILLEN 1968; LARK 1968). The early data on DNA methylation in the cell cycle of eukaryotes are unclear and controversial. It was supposed that in sea urchin embryo DNA is not methylated until the gastrula stage (COMB 1965), but it was later established that DNA methylation proceeds during all developmental stages of the sea urchin, even the earliest ones (POLLOCK et al. 1978; BIRD et al. 1979).

It has been suggested that an essential time gap exists between the replication and methylation of DNA in mouse fibroblasts (BURDON and ADAMS 1969), during the meiotic cycle in the lily (HOTTA and HECHT 1971) and during mitosis in slime mold (EVANS et al. 1973). It has been noted that in mouse fibroblasts methylation of DNA may be completed only a few hours after its synthesis (ADAMS 1971), and a newly synthesized DNA is not methylated (ADAMS 1974). KAPPLER (1970) suggested that the time gap between replication and modification of DNA in cultured mouse adrenal cells does not exceed 2 min. On the other hand, DNA methylation in synchronized HeLa cells continues for at least 30 min after synthesis is over (GERACI et al. 1974). In normal or transformed BHK 21 fibroblasts DNA methylation occurred at the end of the S phase but ceased at the G_2 phase (RUBERY and NEWTON 1973). In "Crow" cells derived from human retro-orbital hemangioma and in cultured Chinese hamster ovary cells DNA methylation is delayed for several hours after strand synthesis, but this delayed methylation is completed before the DNA strand acts as a template for DNA replication at the next S phase (WOODCOCK et al. 1982). The essential DNA methylation in cells of *Euglena gracilis* proceeds during S phase and mitosis, with a minimal incorporation of CH_3 groups during G_2 phase (VALENCIA and BERTAUX 1974). Methylation of DNA in synchronized BHK 21 HS 5 fibroblasts followed a biphasic pattern with maximal methyl incorporations during both S phase and mitosis (BUGLER et al. 1980). The rapidly reannealing DNA synthesized in the Chinese hamster cells during S phase is enriched with 5mC; it is believed that the replication initiation sites contain highly repeated and heavily methylated sequences (SCHNEIDERMANN and BILLEN 1973). It has been suggested that two types of DNA methylases are present in the HeLa cell during S phase; one is responsible for GC-rich DNA methylation and acts at the early S phase, while the other methylates the AT-rich sequences during the late S phase (VOLPE and EREMENKO 1978). In mouse fibroblasts a significant mitotic DNA methylation takes place without any DNA synthesis de novo in the presence of hydroxyurea, added either during mitosis or G_1 phase of the first cell cycle (BUGLER et al. 1980). In mouse 3T3 cells incubated in a medium without methionine DNA synthesis proceeds for a few hours, and the DNA synthesized is essentially (by about 40%) undermethylated (CULP and BLACK 1971). Thus, in animal cells the synthesis of DNA under some conditions may proceed without methylation.

Glucocorticosteroid hormones, as well as other gene inductors, may induce in rat liver DNA methylation which is not associated with replication (VANYUSHIN et al. 1973b; ROMANOV et al. 1976; ROMANENKO et al. 1979; VANYUSHIN and ROMANENKO 1979). This induced DNA methylation is reversible and affects repetitive but not unique sequences (ROMANOV and VANYUSHIN 1981). Noticeable DNA methylation proceeds during repair synthesis in nonproliferating human lymphocytes (DRAHOVSKY et al. 1976). We have detected an increase in the degree of DNA methylation in rat brain neurons during the formation of conditional reflex; this DNA methylation induced by training is not associated with replication (GUSKOVA et al. 1977; VANYUSHIN et al. 1977).

Thus, in eukaryotes DNA may be methylated on, as well as after, replication. One should therefore descriminate between replicative and postreplicative DNA methylation. These DNA modifications may prove to be different with regard to the nature and availability of sites accepting CH_3 groups in DNA of chromatin, and they seem to be carried out by different DNA methylases.

5   Replicative DNA Methylation

DNA replication in animals and plants proceeds discontinuously through formation and subsequent ligation of short (about 5S) Okazaki fragments. This raises the following questions: (a) Are the Okazaki fragments and the ligation stretches methylated? (b) What are the degree and specificity of methylation in the respective replication intermediates?

5.1   Two Stages: Degree of Methylation of Replication Intermediates

The DNA replication pattern strongly depends on cell concentration (density) in a medium (KIRYANOV et al. 1980). When the cell density in a layer (L cells) or in a liquid medium (tobacco cell suspension culture) is not high ($1–2 \times 10^5$ cells per cm^2 or per ml respectively), [^{3}H]thymidine quickly incorporates into the newly formed DNA fragments of $> 5S$; at high cell concentrations ($4–5 \times 10^5/cm^2$) [^{3}H]thymidine incorporation into DNA is almost uninhibited, but ligation is very strongly suppressed and for a long period (about 1 h) the newly synthesized DNA is represented by Okazaki fragments (DEMIDKINA et al. 1979; BASKITE et al. 1980; KIRYANOV et al. 1980). This indicates that the second DNA replication stage, ligation, is selectively sensitive to some physiological factors, and blocking of the Okazaki fragment ligation may be an early stage of realization of the known phenomenon – contact inhibition of cell division. Thus, in animals and plants there is a certain similar mechanism for regulation of the Okazaki fragment ligation on DNA replication. This was used for preparation of large amounts of Okazaki fragments from animal (KIRYANOV et al. 1980) and plant (BASHKITE et al. 1980) cells. The patterns of DNA radioactivity in alkaline sucrose gradients after incubation of the cells with [^{3}H]thymidine or with [methyl-

Table 1. Methylation degree of the radioactive, newly synthesized DNA in transformed mouse fibroblast (L cells) and tobacco cell suspension cultures

Growth conditions	Duration of incubation with [6-^{3}H]uridine	100·5mC/(C+5mC)		
		DNA replication intermediates		Total DNA synthesized
		$\leqq$5S	>5–6S	
L cells (mouse fibroblasts)				
2×10^5 cells/cm^2	15 min, 1–5 h	2.8±0.2	4.2±0.1	
2×10^5 cells/cm^2	24 h	2.7±0.3	4.2±0.1	4.4±0.1
4×10^5 cells/cm^2	24 h	2.8±0.1	4.1±0.1	4.3±0.2
2×10^5 cells/cm^2, 250 µmol SIBA	15 min	2.8±0.2		2.9±0.2
2×10^5 cells/cm^2, 250 µmol SIBA	4 h			2.8±0.2
2×10^5 cells/cm^2, 10 µg/ml cyclo-heximide for 2–3 h	1 h			6.0–6.2
Tobacco cells				
4×10^5 cells/ml	4 h	17.0±0.4	40.2±0.3	
4×10^5 cells/1 mg/l 2,4-D	4 h	16.2±0.6	38.1±0.5	
4×10^5 cells/5 mg/l 2,4 D	4 h	20.2±0.6	20.1±0.5	

SIBA, 5′-deoxy-5′-S-isobutylthioadenosine; 2,4 D, 2,4-dichlorophenoxyacetic acid. From: BASHKITE et al. (1980); KIRYANOV et al. (1980); KIRYANOV et al. (1982)

^{3}H]methionine are very similar (KIRYANOV et al. 1980). Thus, in general, the DNA replication and methylation patterns fit well. It is noteworthy that Okazaki fragments are methylated before being ligated, as soon as they have been formed in the replication fork, and independently of the time of inhibition of ligation. The inhibition of ligation does not influence the methylation level of Okazaki fragments (Table 1).

Radioactivity from [methyl-^{3}H]methionine was detected not only in Okazaki fragments but also in shorter (<5S) replication products (KIRYANOV et al. 1980). This appears most clearly on the strongly inhibited (by 98%) DNA synthesis in L cells (4×10^5/cm^2) incubated for 30 min with 1-β-D-arabinofuranosylcytosine (10 µg/ml). Under these conditions, a newly formed DNA is represented mainly by a few short ($\leqq$5S) methylated fragments (DEMIDKINA et al. 1979). Thus, in animal cells replicative DNA methylation seems to start as soon as the shortest (even <5S) DNA fragment is available for DNA methylase in the replication fork. Therefore, this particular step of DNA methylation must be carried out by an enzyme which is a constituent of the replication complex.

For the quantitative determination of the methylation degree (MD) in DNA, the cells were grown with labeled DNA precursors ([2-^{14}C]uridine, [6-^{3}H]uridine, or [2-^{14}C]orotic acid). Isolated DNA was hydrolyzed into bases and the radioactivity of cytosine and 5mC separated was measured. The methylation degree (MD = $100 \times 5mC/C + 5mC$) of Okazaki fragments measured in this way is about 1.5 times lower in L cells and 2.3 times lower in tobacco cells than in total nuclear DNA (Table 1). At the same time, the longer ($>$ 5S) replication intermediates are methylated additionally: their methylation degree corresponds well to that of mature or total nuclear DNA (Table 1).

Thus, in animals and plants there are at least two stages of replicative DNA methylation, methylation of Okazaki fragments and methylation of ligated replication intermediates.

Now a new question arises: What is methylated in a ligated DNA chain at the second stage of replicative methylation, Okazaki fragments already made and partially methylated, or the ligation stretches newly formed, or both?

To decide between these alternatives, L cells were grown up to a density of 3×10^5 cells/cm^2 (at this cell density, synthesis of Okazaki fragments proceeds without their ligation). Cells were then incubated for 30 min or 4 h with [6-^{3}H]uridine, transferred to a new medium without the label, and grown for several days. DNA synthesized during a 30-min incubation with the label was represented by Okazaki fragments (MD = 2.9) only; DNA formed by a 4-h incubation, mainly by long ($\geq$ 13S) intermediates (MD = 4.0–4.6) (KIRYANOV et al. 1982). After growing these cells for 1–5 days in a nonradioactive medium we could follow the methylation fate of different labeled DNA replication intermediates which appear now as constituents of mature DNA in a few cell populations. It has been shown that no essential postreplicative methylation of long replication intermediates occurs (KIR-YANOV et al. 1982). Thus, replicative DNA methylation terminates in the S phase and outside of this phase there are no essential changes in DNA methylation in L cells. This agrees with the suggestion that DNA methylation proceeds during the S phase and should be finished before the subsequent S phase (WOODCOCK et al. 1982). The radioactive DNA methylation level in cells incubated for 30 min with [6-^{3}H]uridine does not change during the subsequent growth of the cells in a medium without the label for at least 5 days and is equal to the MD of Okazaki fragments (KIRYANOV et al. 1982).

Thus, the Okazaki fragments are methylated only during the first stage of replicative DNA methylation, and there is essentially no additional methylation in the cell cycle. The second stage of replicative DNA methylation concerns only the ligation sites (sequences) formed between Okazaki fragments.

5.2 DNA Methylation in the Presence of Various Inhibitors

In L cells incubated with 250–500 µmol 5′-deoxy-5′-S-isobutyl-thioadenosine (SIBA) DNA replication is inhibited by 60%–80%, but the kinetics of the Okazaki fragment ligation and their MD of 2.9 do not change (KIR-YANOV et al. 1982; Table 1). The MD of high-molecular-weight DNA synthesized in the cells incubated even for 4 h with [6-^{3}H]uridine in the presence of SIBA remains the same as that of Okazaki fragments (2.75), while in DNA of the control (grown without SIBA) cells MD is normal, equal to

4.1 (KIRYANOV et al. 1982). One may therefore conclude that SIBA does not affect the methylation of Okazaki fragments and strongly inhibits methylation of the DNA ligation stretches.

Auxin (2,4-dichlorophenoxyacetic acid), at a concentration of 5 mg/l, does not affect the ratios between replication fragments or methylation of Okazaki pieces in tobacco cell suspension culture, but inhibits the second step of replicative DNA methylation, i.e., the methylation of linked fragments (Table 1; BASHKITE et al. 1980).

Thus, methylation of Okazaki fragments is quite conservative – it depends very little on the cell density in a medium, and it is not inhibited by the competitive inhibitor of methylation reactions in the cell (SIBA) or by auxin. SIBA and auxin effectively control (inhibit) methylation of ligation stretches in newly formed DNA.

The DNA methylation pattern strongly depends on chromatin organization and the availability of methylation sites in it; for example, when methylated in vitro in the presence of heterologous bacterial DNA methylases, rat liver chromatin DNA accepts markedly fewer CH_3 groups than the same – but naked – DNA (KIRYANOV et al. 1981). Since the synthesis of DNA and histones is coordinated and the pool of histones in the cell is relatively small (SEALE 1976), the inhibition of histone synthesis may result in the formation of chromatin with additional DNA methylation sites exposed to DNA methylases. The S phase that began in 3T3 cells in the presence of cycloheximide proceeds up to the end, and chromatin in these cells contains about two times fewer histones than the chromatin formed under normal growth conditions (SEALE 1976). In the L cells incubated in the presence of 10–100 µg/ml cycloheximide, protein synthesis is totally inhibited, but DNA synthesis by only 80%; under these conditions DNA synthesized during 1–2 h of cell incubation with [6-^{3}H]uridine is very heavily methylated (MD = 6.0–6.2) (KIRYANOV et al. 1982). This means that some of the CG sequences in DNA of newly formed chromatin (nucleosomes) are normally prevented by histones from being methylated.

5.3 A General Scheme

Replicative DNA methylation in animals and higher plants proceeds in two stages; first Okazaki fragments and second, the ligation stretches are methylated. Since Okazaki fragments synthesized during inhibition of their ligation (at high cell density) have been found in nucleosomes (KIRYANOV et al. 1982), this DNA intermediate may be a constituent of a nucleosome which is supposed to be an elementary chromatin replication unit in eukaryotes. Okazaki fragment, after being methylated at one stage, forms a nucleosome, and seems therefore to be unable to be further methylated in a cell cycle. We know that Okazaki fragments are undermethylated compared with total DNA, and their MD in L cells is 2.9. Taking into account that the total DNA MD in these cells is about 4.2, the lengths of an Okazaki fragment and of a ligation stretch are about 140–145 and 60 nucleotides respectively; one may calculate that the methylation level of the ligation

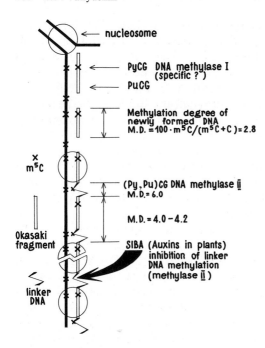

Fig. 1. Replicative DNA methylation in animal and plant cells. Numbers represent data on methylation degree (*MD*) of newly synthesized DNA in transformed mouse fibroblasts (L cells)

area should be at least twice (about 6.0) that of Okazaki fragments. This means that all CG sequences in the ligation stretches are fully methylated (nonrandom CG distribution between nucleosomal and internucleosomal DNA is ruled out). It follows from these data that Okazaki fragment contains about two CG sites, of which only one is methylated; in a ligation stretch there is about one CG site, and it is methylated. It seems that ligation of Okazaki fragments occurs mainly after the formation of nucleosomes, and as a matter of fact, the second step of replicative DNA methylation may deal with internucleosomal (ligation) DNA stretches. As is well known, internucleosomal DNA is more available for interactions with various proteins, for example, with methylases (KIRYANOV et al. 1981) or hormone receptors, than is nucleosomal DNA. Besides, in animal cells DNA methylase has been found to be localized predominantly in the internucleosomal regions of condensed chromatin (CREUSOT and CRISTMAN 1981). All this may explain at least partially, why ligation stretches, which may correspond to internucleosomal DNA, are so highly methylated.

In the model for replicative DNA methylation in eukaryotes that we suggest (Fig. 1), we propose that in a nucleus there must be at least two different DNA methylases (or two different forms of one enzyme) which are different in the recognition sites and in the way they function at the various stages of DNA replication (KIRYANOV et al. 1982). It has been shown that the 5mC distribution among pyrimidine isostichs in Okazaki fragments and mature or total nuclear DNA of L cells is very different: Pu-C-Pu sequences in Okazaki fragments are much less methylated than those in total

DNA (DEMIDKINA et al. 1979). Therefore, one may propose that at the first stage of replicative methylation, DNA methylase I predominantly methylates Py-CG but not Pu-CG sequences. As a result, in Okazaki fragments only about half of all the CG sites will be methylated (which proved to be the case). It is possible that this DNA methylase I recognizes only hemimethylated sites. Undermethylated Okazaki fragments, then, form nucleosomes and are not available for further methylation. At the second replication stage, DNA methylase II starts to methylate ligation stretches; this enzyme seems to be less specific in the recognition sites, and methylates all CG sequences. DNA methylase II seems to be responsible for total (MD up to 6.0) methylation of CG sites in mouse DNA (including undermethylated ones in Okazaki fragments) during cycloheximide inhibition of histone synthesis and of nucleosome formation.

A possible existence of two DNA methylases in a nucleus is also favored by the data on different sensitivity of the two stages of replicative DNA methylation to SIBA and auxins, as compared with methylation of ligated DNA intermediates. A higher resistance of the Okazaki fragment methylation to SIBA may be associated with a lower affinity of DNA methylase I than of DNA methylase II to this common methylation inhibitor.

It has been suggested that there may be two types of DNA methylases in eukaryotes: (a) a maintaince enzyme which methylates only half-methylated sites, and (b) an initiating enzyme which is able to methylate fully unmethylated DNA (RAZIN and RIGGS 1980). From my point of view, DNA methylase I may be a good candidate for maintaining an enzyme, and DNA methylase II for initiating one; DNA methylase II is able to methylate any CG site.

Thus, the methylation pattern of DNA in eukaryotes depends on the specificity of action of methylases and on the structural organization of chromatin in a nucleus.

6 Gene Activation and Cell Differentiation

The idea that DNA methylation participates in regulation of transcription and cell differentiation (SCARANO et al. 1967; BERDYSHEV et al. 1967; VANYUSHIN et al. 1970; RIGGS 1975; HOLLIDAY and PUGH 1975) has proved to be true and very stimulating. It has been supported by the discovery of tissue and age differences in DNA methylation in animals and plants, of changes in DNA methylation during various functional states of the organism, and of symmetry and heritability of eukaryotic DNA methylation (BIRD 1978; WIGLER 1981; for reviews see RAZIN and RIGGS 1980; DOERFLER 1981; EHRLICH and WANG 1981). A reverse correlation exists between the degree of DNA methylation and transcription. Induced demethylation of DNA results in the activation of transcription, so the methyl groups in regulatory DNA sequences seem to be some sort of signal for negative, or sometimes positive, control of transcription.

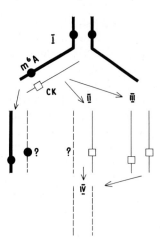

Fig. 2. Formation of unmethylated adenine sites in DNA on replication as a result of cytokinin (*CK*) incorporation

One may suppose that in the chromatin of the interphase nucleus genes exists in three different states: unactive, activated, and active. It seems that controlled replicative undermethylation of cytosine residues detected in the ligation stretches of newly synthesized DNA is an effective control of transcription. As mentioned above, these (internucleosomal) sites seem to be most available to regulatory proteins. This induced DNA undermethylation may be a main mechanism for regulation of cell differentiation in plants by auxins.

I suspect that other natural plant growth regulators, e.g., cytokinins (N^6-substituted adenine derivatives), may effectively influence replicative methylation of adenine residues in DNA of eukaryotes which are known to contain 6mA (protozoa, fungi, algae, higher plants, and probably some invertebrates). The presence of various radicals at the N^6-position in cytokinins should be a barrier for methylation at the same position. If cytokinin appears to be incorporated in the nucleotide sequence recognized by adenine methylase, methylation of the adenine residue with formation of 6mA in this particular sequence will not occur (Fig. 2, *I*). Unfortunately, the nature of this sequence is not known. If it is symmetrical, and a methyl group in one DNA chain is a signal for methylation in the opposite one, a fully (in both chains) unmethylated adenine site may appear in this DNA on replication (Fig. 2, *II*). Besides, it cannot be ruled out that a fully unmethylated adenine site may originate as a result of the cytokinin incorporation in both complementary DNA chains on subsequent replication (Fig. 2, *III*). In any case, from these cytokinin-containing molecules DNA with some fully unmethylated adenine sites may originate (Fig. 2, *IV*). After incubation of wheat seedlings with 6-benzylaminopurine, the 6mA amount in their DNA has diminished (Vanyushin, unpublished). Thus, incorporation into DNA and induction of the formation of unmethylated adenine sites in it may possibly be a mechanism for the regulation of replication, transcription, and cell differentiation by cytokinins.

7 Summary

Replicative DNA methylation in eukaryotes proceeds in at least two stages – first, Okazaki fragments, and, second, the DNA ligation stretches are methylated. These two stages seem to be carried out by different DNA methylases. Okazaki fragments are methylated only at one stage and remain undermethylated in mature DNA. Methylation of Okazaki fragments does not depend on cell concentration in a medium and is resistant to S-isobutyl-thioadenosine in animal cells and auxin in plant cells. These compounds selectively inhibit methylation of DNA ligation stretches. Replicative under-methylation of ligation stretches in newly formed DNA regulated by auxins may be one of the mechanisms for control of cell differentiation in plants. Kinins seem to regulate gene expression by inhibition of methylation of some adenine residues in eukaryotic DNA.

References

Adams RLP (1971) The relationship between synthesis and methylation of DNA in mouse fibroblasts. Biochim Biophys Acta 254:205–212

Adams RLP (1974) Newly synthesized DNA is not methylated. Biochim Biophys Acta 335:365–373

Bashkite EA, Kirnos MD, Kiryanov GI, Alexandrushkina NI, Vanyushin BF (1980) Replication and methylation of DNA in cells of tobacco suspension culture and the effect of auxin. (in Russian) Biokhimiya 45:1448–1456

Berdyshev GD, Korotaev GK, Boyarskikh GV, Vanyushin BF (1967) Nucleotide composition of DNA and RNA from somatic tissues of humpback and its changes during spawning. (in Russian) Biokhimiya 32:988–993

Billen D (1968) Methylation of the bacterial chromosome: an event at the "replication point"? J Mol Biol 31:477–486

Bird A (1978) Use of restriction enzymes to study eukaryotic DNA methylation. II. The symmetry of methylated sites supports semiconservative copying of the methylation pattern. J Mol Biol 118:49–60

Bird A, Taggart M, Smith BA (1979) Methylated and unmethylated DNA compartments in the sea urchin genome. Cell 17:889–901

Bird A, Taggart M, Macleod D (1981) Loss of rDNA methylation accompanies the onset of ribosomal gene activity in early development of $X. laevis$. Cell 26:381–390

Bugler B, Bertaux O, Valencia R (1980) Nucleic acids methylation of synchronized BHK 21 HS5 fibroblasts during mitotic phase. J Cell Physiol 103:149–157

Burdon RH, Adams RLP (1969) The in vivo methylation of DNA in mouse fibroblasts. Biochim Biophys Acta 174:322–329

Burtseva NN, Demidkina NP, Azizov YM, Vanyushin BF (1978) Changes in specificity of DNA methylation in cattle blood lymphocytes under chronic lympholeucosis. (in Russian) Biokhimiya 43:2082–2091

Buryanov YI, Ilyin AV, Skryabin GK (1970) On detection of 6-methylaminopurine in DNA of fungus $Mucor\ hiemalis$ (in Russian). Dokl Akad Nauk SSSR 195:728–730

Buryanov YI, Eroshina NV, Vagabova LM, Ilyin AV (1972) Detection of 6-methylaminopurine in DNA of the higher plant pollen. (in Russian) Dokl Akad Nauk SSSR 205:700–703

Chvojka LA, Sulimova GE, Bulgakov R, Bashkite EA, Vanyushin BF (1978) Changes in the content of 5-methylcytosine in plant DNA due to flowering gradient. (in Russian) Biokhimiya 43:996–1000

Comb DG (1965) Methylation of nucleic acids during sea urchin embryo development. J Mol Biol 11:851–855

Creusot F, Cristman JK (1981) Localization of DNA methyltransferase in the chromatin of Friend erythroleukemia cells. Nucleic Acids Res 9:5359–5381

Culp LA, Black PH (1971) DNA synthesis in normal and virus-transformed mammalian cells after methionine deprivation. Biochim Biophys Acta 247:220–232

Cummings DJ, Tait A, Goddard JM (1974) Methylated bases in *Paramecium aurelia*. Biochim Biophys Acta 374:1–11

Demidkina NP, Kiryanov GI, Vanyushin BF (1979) Methylation of newly synthesized DNA in mouse fibroblast culture. (in Russian) Biokhimiya 44:1416–1426

Deumling B (1981) Sequence arrangement of highly methylated satellite DNA of a plant, *Scilla*: a tandemly repeated inverted repeat. Proc Natl Acad Sci USA 78:338–342

Doerfler W (1981) DNA methylation – a regulatory signal in eukaryotic gene expression. J Gen Virol 57:1–20

Doheim M, Sulimova GE, Vanyushin BF (1978) Distribution of 5-methylcytosine in pyrimidine oligonucleotides of higher plant DNAs. (in Russian) Biokhimiya 43:1312–1318

Drahovsky D, Lacko I, Wacker A (1976) Enzymatic DNA methylation during repair synthesis in nonproliferating human peripheral lymphocytes. Biochim Biophys Acta 447:139–143

Ehrlich M, Wang RY-H (1981) 5-Methylcytosine in eukaryotic DNA. Science 219:1350–1357

Evans HH, Evans TE, Littman S (1973) Methylation of parental and progeny DNA strands in *Physarum polycephalum*. J Mol Biol 74:563–572

Geraci D, Eremenko T, Cocchiara R, Giranieri A, Scarano E, Volpe P (1974) Correlation between synthesis and methylation of DNA in HeLa cells. Biochem Biophys Res Communs 57:353–357

Gorovsky MA, Hattman S, Pleger GL (1973) N^6-Methyladenine in the nuclear DNA of a eukaryote, *Tetrahymena pyriformis*. J Cell Biol 56:697–701

Gruenbaum Y, Naveh-Many T, Cedar H (1981) Sequence specificity of methylation in higher plant DNA. Nature 292:860–862

Guskova LV, Burtseva NN, Tushmalova NA, Vanyushin BF (1977) The DNA methylation level in nuclei of neurons and glia from the rat brain hemisphere and its changes during the conditional reflex formation. (in Russian) Dokl Akad Nauk SSSR 233:993–996

Harbers K, Harbers B, Spencer JH (1975) Nucleotide clusters in DNAs. XII. The distribution of 5-methylcytosine in pyrimidine oligonucleotides of mouse L-cell satellite DNA and mainband DNA. Biochem Biophys Res Communs 66:738–746

Hattman S, Kenney C, Berger L, Pratt K (1978) Comparative study of DNA methylation in three unicellular eukaryotes. J Bacteriol 135:1156–1157

Holliday R, Pugh JE (1975) DNA modification mechanisms and gene activity during development. Science 187:226–232

Hotta Y, Hecht N (1971) Methylation of *Lilium* DNA during meiotic cycle. Biochim Biophys Acta 238:50–56

Kappler JW (1970) The kinetics of DNA methylation in cultures of a mouse adrenal cell line. J Cell Physiol 75:21–32

Kappler JW (1971) The 5-methylcytosine content of DNA: Tissue specificity. J Cell Physiol 78:33–36

Kirnos MD, Merkulova NA, Borkhsenius SN, Vanyushin BF (1980) Character of the macronucleus DNA methylation in protozoan *Tetrahymena pyriformis*. (in Russian) Dokl Akad Nauk SSSR 255:225–227

Kirnos MD, Alexandrushkina NI, Vanyushin BF (1981) 5-Methylcytosine in pyrimidine sequences of plant and animal DNA: specificity of DNA methylation. (in Russian) Biokhimiya 46:1458–1474

Kiryanov GI, Kirnos MD, Demidkina NP, Alexandrushkina NI, Vanyushin BF (1980) Methylation of DNA in L cells on replication. FEBS Lett 112:225–228

Kiryanov GI, Smirnova TA, Isaeva LV, Vanyushin BF, Buryanov YI (1981) Limited accessibility of DNA methylation sites for bacterial methylases M. *Eco*RII and M. *Eco dam* in chromatin at different levels of organization. (in Russian) Biokhimiya 46:1887–1895

Kiryanov GI, Isaeva LV, Kirnos MD, Ganicheva NI, Vanyushin BF (1982) Replicative methylation of DNA in L cells: effect of S-isobutyladenosine and cycloheximide and possible existence of two DNA methylases. (in Russian) Biokhimiya 47:153–161

Kudryashova IB, Vanyushin BF (1976) In vitro methylation of nuclear DNA from various rat organs: tissue and age differences in DNA acceptor ability. (in Russian) Biokhimiya 41:1106–1115

Lark C (1968) Effect of methionine analogs, ethionine and norleucine, on DNA synthesis in *Escherichia coli*15T⁻. J Mol Biol 31:401–414

Manes C, Menzel P (1981) Demethylation of CpG sites in DNA of early rabbit trophoblasts. Nature 293:589–590

McGhee JD, Ginder GD (1979) Specific DNA methylation sites in the vicinity of the chicken β-globin gene. Nature 280:419–420

Pakhomova MV, Zaitseva GN, Belozersky AN (1968) The presence of 5-methylcytosine and 6-methylaminopurine in DNA of some algae. (in Russian) Dokl Akad Nauk SSSR 182:712–715

Pollock JM, Swihart M, Taylor H (1978) Methylation of DNA in early development: 5-methyl-cytosine content of DNA in sea urchin sperm and embryos. Nucleic Acids Res 5:4855–4863

Rae PMM (1976) Hydroxymethyluracil in eukaryote DNA: a natural feature of the Pyrrophyta (dinoflagellates). Science 194:1062–1064

Razin A, Riggs AD (1980) DNA methylation and gene function. Science 210:604–610

Riggs AD (1975) X inactivation, differentiation and DNA methylation. Cytogenet Cell Genet 14:9–25

Romanenko EB, Palmina NP, Vanyushin BF (1979) Correlation of the increase in DNA methylation and antioxidant activity of mouse liver nuclear lipids after administration of antioxidant, and in Ehrlich ascite carcinoma. (in Russian) Biokhimiya 44:1754–1761

Romanenko EB, Obukhova LK, Vanyushin BF (1981) Changes in DNA methylation in mice with age and under the influence of hydrocortisone and antioxidant. (in Russian) Biologicheskie nauki 2:63–70

Romanov GA, Vanyushin BF (1981) Methylation of reiterated sequences in mammalian DNAs. Effects of the tissue type, age, malignancy and hormonal induction. Biochim Biophys Acta 653:204–218

Romanov GA, Kiryanov GI, Dvorkin VM, Vanyushin BF (1976) Effects of hydrocortisone on methylation and molecular population of DNA in rat liver. (in Russian) Biokhimiya 41:1038–1043

Rubery EP, Newton AA (1973) DNA methylation in normal and tumour virus-transformed cells in tissue culture. The level of DNA methylation in BHK 21 cells transformed by Py virus. Biochim Biophys Acta 324:24–36

Salomon R, Kaye AM (1970) Methylation of mouse DNA in vivo: di- and tripyrimidine sequences containing 5-methylcytosine. Biochim Biophys Acta 204:340–351

Scarano E, Iaccarino M, Grippo P, Parisi E (1967) The heterogeneity of thymine methyl group origin in DNA pyrimidine isostichs of developing sea urchin embryos. Proc Natl Acad Sci USA 57:1394–1400

Schneidermann MH, Billen D (1973) Methylation of rapidly reannealing DNA during the cell cycle of Chinese hamster cells. Biochim Biophys Acta 308:352–360

Seale RL (1976) Temporal relationship of chromatin protein synthesis, DNA synthesis and assembly of deoxyribonucleoprotein. Proc Natl Acad Sci USA 73:2270–2274

Sneider TW (1972) Methylation of mammalian DNA. Terminal versus internal location of 5-methylcytosine in oligodeoxyribonucleotides from Novikoff hepatoma cell DNA. J Biol Chem 247:2872–2875

Sneider TW (1980) The 5'-cytosine in CCGG is methylated in two eukaryotic DNAs and *Msp*I is sensitive to methylation at this site. Nucleic Acids Res 8:3829–3840

Sobieski DA, Eden FC (1981) Clustering and methylation of repeated DNA: persistence in avian development and evolution. Nucleic Acids Res 9:6001–6015

Sulimova GE, Drozhdenyuk AP, Vanyushin BF (1978) Changes in methylated sequences and molecular population of wheat DNA on germination. (in Russian) Molekularnaya Biologiya 12:496–504

Valencia R, Bertaux O (1974) Les méthylations des acides nucléiques au cours du cycle cellulaire chez *Euglena*. Coll Int CNRS 240:69–75

Van der Ploeg LHT, Flavell RA (1980) DNA methylation in the human γδβ-globin locus in erythroid and nonerythroid tissues. Cell 19:947–958

Vanyushin BF, Belozersky AN (1959) Nucleotide composition of higher plant DNAs. (in Russian) Dokl Akad Nauk SSSR 129:944–946

Vanyushin BF, Romanenko EB (1979) Changes in DNA methylation in rat during ontogenesis and under effects of hydrocortisone. (in Russian) Biokhimiya 44:78–85

Vanyushin BF, Masharina LV, Belozersky AN (1962) On distribution of pyrimidines in DNA. (in Russian) Dokl Akad Nauk SSSR 147:958–961

Vanyushin BF, Belozersky AN, Kokurina NA, Kadirova DX (1968) 5-Methylcytosine and 6-methylaminopurine in bacterial DNA. Nature 218:1066–1067

Vanyushin BF, Tkacheva SG, Belozersky AN (1970) Rare bases in animal DNA. Nature 225:948–949

Vanyushin BF, Kadyrova DK, Karimov KK, Belozersky AN (1971) Minor bases in higher plant DNA. (in Russian) Biokhimiya 36:1251–1258

Vanyushin BF, Mazin AL, Vasilyev VK, Belozersky AN (1973a) The content of 5-methylcytosine in animal DNA: the species and tissue specificity. Biochim Biophys Acta 299:397–403

Vanyushin BF, Nemirovsky LE, Klimenko VV, Vasilyev VK, Belozersky AN (1973b) The 5-methylcytosine in DNA of rats. Tissue and age specificity and the changes induced by hydrocortisone and other agents. Gerontologia 19:138–152

Vanyushin BF, Tushmalova NA, Guskova LV, Demidkina NP, Nikandrova LP (1977) Changes in the DNA methylation level in rat brain during conditional reflex formation. (in Russian) Molekularnaya Biologiya 11:181–187

Vanyushin BF, Bashkite EA, Fridrich A, Chvojka L (1981) DNA methylation in wheat seedlings and influence of phytohormones. (in Russian) Biokhimiya 46:47–54

Volpe P, Eremenko T (1978) A language of DNA modification during its replication. Proc of the XIV Int Congress of Genetics, part 1. Moscow, p 194

Waalwijk C, Flavell RA (1978) DNA methylation at a CCGG sequence in the large intron of the rabbit β-globin gene: tissue-specific variations. Nucleic Acids Res 5:4631–4641

Wigler MH (1981) The inheritance of methylation patterns in vertebrates. Cell 24:285–286

Woodcock DM, Adams JK, Cooper IA (1982) Characteristics of enzymatic DNA methylation in cultured cells of human and hamster origin, and the effect of DNA replication inhibition. Biochim Biophys Acta 696:15–22

Wyatt GR (1951) The purine and pyrimidine composition of deoxypentose nucleic acids. Biochem J 48:584–590

Zinkovskaya GG, Berdyshev GD, Vanyushin BF (1978) Tissue-specific decrease and changes in the DNA methylation character in cattle with aging. (in Russian) Biokhimiya 43:1883–1892

5-Azacytidine, DNA Methylation, and Differentiation *

S.M. Taylor[1], P.A. Constantinides[2], and P.A. Jones[1,3]

1 Introduction

First synthesized by PISKALA and SORM (1964), 5-azacytidine (5-aza-CR) is an S-triazine nucleoside analogue of cytidine which differs from cytidine only by the replacement of the 5-carbon atom with a nitrogen. The ring is susceptible to spontaneous hydrolysis under neutral or mildly alkaline conditions, giving rise to hydrolysis products with a variety of effects in living systems (CIHAK and SORM 1965; PITHOVA et al. 1965; NOTARI and DE YOUNG 1975). The drug is an effective chemotherapeutic agent in the treatment of human acute myelogenous leukemia and has been studied in a variety of in vivo and in vitro systems (VESELY and CIHAK 1978). Considerable interest has also been aroused in the use of the analogue as a tool for the study of differentiation and in its potential to induce the expression of inactive genes. Recent studies on the induction of new gene expression by 5-aza-CR and other cytidine analogues modified in the 5 position will be reviewed here.

* Supported by grant GM 30892 from the National Institute of General Medical Sciences

1 Division of Hematology-Oncology, Childrens Hospital of Los Angeles, University of Southern California, 4650 Sunset Boulevard, Los Angeles, CA 90027, USA
2 Department of Medical Biochemistry, University of Cape Town, Observatory 7925, South Africa
3 Departments of Pediatrics and Biochemistry, University of Southern California, 4650 Sunset Boulevard, Los Angeles, CA 90027, USA

Current Topics in Microbiology and Immunology, Vol. 108
© Springer-Verlag Berlin·Heidelberg 1984

2 Phosphorylation of Cytidine Analogues

Uridine-cytidine kinase (ATP:uridine 5′-phosphotransferase, EC 2.7.1.48), catalyzes the phosphorylation of uridine, cytidine, and a number of their analogues to nucleoside monophosphates in the presence of ATP and Mg^{++} (CAPUTTO 1962; ANDERSON 1973). It is important in the salvage pathway for pyrimidines and is especially active in rapidly proliferating cells (SKOLD 1960; LUCAS 1967), since phosphorylation of pyrimidine nucleosides is the rate-limiting step in the production of nucleotides (ANDERSON and BROCK-MAN 1964).

Phosphorylation of 5-aza-CR by uridine-cytidine kinase has been studied extensively (VESELY and CIHAK 1978, for review). LIACOURAS and ANDERSON (1979) showed that 5-aza-CR, cytidine, and uridine are all phosphorylated by a single enzyme and not by separate, closely related enzymes. The apparent K_m of the purified kinase from hepatoma cells for 5-aza-CR was 20 and 120 times higher than that for uridine and cytidine respectively (LIACOURAS and ANDERSON 1979). 5-Aza-CR is a relatively weak competitive inhibitor of cytidine phosphorylation, whereas both cytidine and uridine are potent competitive inhibitors of 5-aza-CR phosphorylation (LEE et al. 1974, 1975; LIACOURAS and ANDERSON 1979). 5,6-Dihydro-5-azacytidine (DH-aza-CR), which is stable under neutral and alkaline conditions (BEISLER et al. 1976), has been compared with 5-aza-CR in a number of systems. Uridine-cytidine kinase prepared from HeLa cells phosphorylated both 5-aza-CR and DH-aza-CR, and their respective apparent K_m values were approximately two and 19 times higher than that of cytidine (FUTTERMAN et al. 1978). Pseudoisocytidine (ψICR), a C-riboside analogue of cytidine with a stable ring structure (CHOU et al. 1979), is also phosphorylated by the kinase, with an apparent K_m approximately 100 times higher than that for cytidine (CHOU et al. 1979).

Deoxycytidine kinase catalyses the phosphorylation of 5-aza-2′-deoxycytidine (5-aza-CdR) with an apparent K_m five times higher than that for deoxycytidine (MOMPARLER and DERSE 1979). Deoxycytidine acts as a potent competitive inhibitor of the phosphorylation of 5-aza-CdR, and can completely eliminate the effects of the analogue in several systems (MOMPARLER and GOODMAN 1977).

3 Inactivation of Cytidine Analogues

Enzymatic deamination of 5-azacytosine analogues occurs both in vitro and in vivo (VESELY and CIHAK 1978). Both 5-aza-CR (CHABNER et al. 1974; STOLLER et al. 1976) and DH-aza-CR (FUTTERMAN et al. 1978; VOYTEK et al. 1977) are substrates for cytidine deaminase (EC 3.5.4.5), and the apparent K_m values indicate that DH-aza-CR binds to the enzyme more effectively than cytidine and ten times better than 5-aza-CR. The 5-azauracil analogues formed by deamination inhibit the de novo synthesis of orotate (VESELY and CIHAK 1978).

5-Azacytidine is also deaminated to form 5-aza-2′-deoxyuridine and other metabolites. These metabolites can be phosphorylated and their formation is inhibited by tetrahydrouridine, which inhibits the deaminase (CIHAK 1978). The deaminated products are believed to interfere with de novo thymidylate synthesis (MALEY and MALEY 1960; VESELY et al. 1969). Similarly, intracellular deamination of 5-fluoro-2′-deoxycytidine results in the formation of FdUrd (DIASIO et al. 1978), a potent inhibitor of thymidylate synthetase.

4 Incorporation into Nucleic Acids

5-Azacytidine is incorporated into both RNA and DNA following reduction by ribonucleotide diphosphate reductase (VESELY and CIHAK 1978, for review). Incorporation into RNA impairs rRNA processing and maturation and the amino acid acceptor function of tRNA (LEE and KARON 1976), perhaps by inhibiting the modification of cytosine residues (LU and RANDERATH 1979). GLAZER et al. (1980) have shown that dH-aza-CR is incorporated into all species of nuclear RNA and CONSTANTINIDES et al. (1980) reported the incorporation of ^{14}C-labeled analog into DNA. High concentrations of dH-aza-CR are required to produce measurable incorporation into DNA, perhaps due to the lower affinity of uridine cytidine kinase for this analogue and rapid catabolism by cytidine deaminase (FUTTERMAN et al. 1978; VOYTEK et al. 1977). The effects of this analogue on the structure and function of nucleic acids have not yet been well documented.

The phosphorylation and incorporation of 5-aza-CdR into nucleic acids occurs primarily in rapidly proliferating tissue, such as calf thymus and spleen (SORM et al. 1966) and mouse leukemic cells (VESELY and CIHAK 1977), where deoxycytidine kinase activity is relatively high. This analogue is incorporated only into DNA, due to the irreversible nature of the reaction catalyzed by ribonucleotide reductase (LI et al. 1970). CHOU et al. (1979) have demonstrated the incorporation of ^{14}C-ψICR into both RNA and DNA, although the rate of incorporation was 100 times lower than that of ^{14}C-cytidine. This has been ascribed to the lower affinity of the kinase for ψICR (CHOU et al. 1979).

Incorporation of high levels of 5-aza-CR into DNA in *Escherichia coli* results in helix instability and disruption of secondary structure (ZADRAZIL et al. 1965). Lability of incorporated 5-azacytosine was suggested as a cause of this instability. Chromatid breakage during G_2 in a hamster fibroblast cell line treated with 1 µg/ml (approximately 50 µM) 5-aza-CR may be due to incorporation of the drug into DNA (KARON and BENEDICT 1972). Similar concentrations of 5-aza-CR inhibit or delay condensation of some chromosome segments corresponding to the G-bands in human lymphocyte cultures (VIEGAS-PEQUIGNOT and DUTRILLAUX 1976). The possibility of ring cleavage of the triazine ring in the polynucleotide cannot be completely ruled out and may be important in some of these effects of 5-aza-CR in biological systems. However, 5-azacytosine can be recovered intact from the DNA of cells treated with 1–10 μM 5-aza-CR (JONES and TAYLOR 1981). In addition, several stable analogues of cytidine (eg ψICR and FCdR) mimic the ability of 5-aza-CR to induce differentiation in cultured cells (JONES and TAYLOR 1980), so that ring cleavage is unlikely to play a role in mediating this effect.

5 5-Azacytidine and Cellular Differentiation

5-Azacytidine has pronounced effects on the stability of the differentiated state of cultured cells. Micromolar concentrations of the analogue induce the formation of fully functional, biochemically differentiated striated muscle cells from the nonmuscle line C3H/10T$^1/_2$ CL8 (10T$^1/_2$) (CONSTANTINIDES et al. 1977, 1978). These muscle cells have elevated levels of myosin ATPase activity and develop functional acetylcholine receptors capable of initiating a twitching response to added acetylcholine. They also twitch spontaneously in culture, and represent normal muscle cells which are never seen in untreated 10T$^1/_2$ cultures or in cultures exposed to a wide variety of other agents.

Biochemically differentiated adipocytes and chondrocytes were also observed in treated cultures, so the changes in differentiation induced by 5-aza-CR are not restricted to the muscle phenotype (TAYLOR and JONES 1979). They are not restricted to cells of the 10T$^1/_2$ line, since functional muscle, fat, and cartilage cells arise in Swiss 3T3 cells exposed to micromolar concentrations of 5-aza-CR (TAYLOR and JONES 1979). The CVP3SC6 line derived from adult mice (NESNOW and HEIDELBERGER 1976) and cloned lines of oncogenically transformed 10T$^1/_2$ cells also form functionally striated muscle after 5-aza-CR treatment (TAYLOR and JONES 1979, 1982a). Since the transformed cell lines are tumorigenic in mice (JONES et al. 1976), 5-aza-CR can induce the expression of normal differentiated phenotypes from tumorigenic cells.

The induction of mesenchymal differentiation by 5-aza-CR in these systems is not due to a selection of cells from the original population with a predisposition to differentiate. Thus, a high proportion of subclones (TAYLOR and JONES 1979) or single 10T$^1/_2$ cells (TAYLOR and JONES 1982a) form muscle and fat colonies after drug treatment. The formation of all three new phenotypes (muscle, fat, and cartilage) has been observed in some cases in the progeny of single cells exposed to 5-aza-CR.

The development of new phenotypes shows an absolute requirement for cell division after treatment, and more than two divisions are required for the expression of the muscle phenotype (TAYLOR and JONES 1982a), with the maximum number of muscle cells arising eight to ten divisions after treatment. The large number of mitoses which occur in the absence of further drug treatment indicates that the analogue is not present in the DNA of cells expressing a new phenotype. This fact distinguishes the effects of 5-aza-CR on differentiation from those observed for the extinction of phenotypes by bromodeoxyuridine (RUTTER et al. 1973). In these cases, the continued presence of the fraudulent base in DNA is often required for its effect on differentiation to be apparent.

Phenotypic conversion of $10T^1/_2$ cells is dependent on the phase of the cell cycle in which the analogue is administered. Cells synchronized by post-confluence inhibition of growth or isoleucine deprivation show the greatest response when treated early in S phase, and virtually no muscle is induced in cultures treated in G_1 phase (CONSTANTINIDES et al. 1978). The cell cycle specificities for muscle and adipocyte conversions do not differ significantly from each other (TAYLOR and JONES 1982a). Differences between early and late S-phase responses cannot be accounted for by an enhanced incorpora-tion of 5-azacytosine into DNA replicated early in S phase (TAYLOR and JONES 1982a). Thus, 5-azacytosine may be incorporated into DNA control-ling the expression of mesenchymal phenotypes replicated early in S phase so that new phenotypes may be expressed subsequently.

The induction of muscle cells from treated $10T^1/_2$ cells by 5-aza-CR can be inhibited by the subsequent application of certain tumor promoters such as phorbol esters (MONDAL and HEIDELBERGER 1980). Since the $10T^1/_2$ cells also respond to two-stage chemical oncogenesis following treatment with chemical carcinogens and tumor promoters (MONDAL et al. 1976), it may be possible to study oncogenesis, tumor promotion, and differentiation in a single cell system.

5-Azacytidine is incorporated into both the RNA and DNA of treated cells (LI et al. 1970), so it was important to distinguish which of these effects is involved in the mechanism of drug action. The deoxy analogue, 5-aza-2′-deoxycytidine (5-aza-CdR), was more active than the ribo analogue in inducing muscle cell formation (CONSTANTINIDES et al. 1978). Since 5-aza-CdR is incorporated only into DNA (LI et al. 1970), a DNA-linked mecha-nism of action is implied. Two stable analogues of cytidine, FCdR and ψICR, both containing modifications in the 5 position, were also effective at muscle induction in $10T^1/_2$ cells (JONES and TAYLOR 1980). Since the only common feature of these analogues is the modification of the 5 position of the cytosine ring, we interpreted the results to suggest that this position must have significance in the control of gene expression. All of the active analogues were found to be potent inhibitors of the methylation of cytosine residues in DNA at biologically effective concentrations, suggesting that the mechanism of action of the analogues was linked to their abilities to inhibit DNA methylation (JONES and TAYLOR 1980).

WILSON et al. (1983) have shown that therapeutic doses of 5-aza-CdR inhibit the methylation of L1210 leukemic cells growing as an ascites tumor in mice. Inhibition of DNA methylation in tumor cells may therefore ac-count for some of the therapeutic effects of 5-aza-CdR.

Although 5-aza-CR has been reported as being mutagenic in viral sys-tems (HALLE 1968; PRINGLE 1970), we have not found it or the other ana-logues mentioned above to be significantly mutagenic in $10T^1/_2$ or V79 cells (LANDOLPH and JONES 1982). The marked effects of the analogues on differentiation are therefore unlikely to be due to mutation.

6 Induction of Differentiation in Other Systems

5-Azacytidine alters the inducibility of the metallothionein gene in W7 mouse thymoma cells (COMPERE and PALMITER 1981). Unlike most other cell lines, these cells do not express the metallothionein gene in the presence of either cadmium or glucocorticoids. However, the gene becomes inducible when W7 cells are treated for a few hours with 5-aza-CR. Restriction analysis showed that analogue treatment changed the methylation pattern near the gene and resulted in a hypomethylation which was correlated with the expressibility of the gene (COMPERE and PALMITER 1981).

The human promyelocytic leukemia cell line HL60 can be induced to differentiate by the nucleoside analogue (BODNER et al. 1981), and Friend erythroleukemia cells synthesize hemoglobin following treatment with 5-aza-CR or 5-aza-CdR (CREUSOT et al. 1982). Multiple new phenotypes, including muscle, cartilage, and adipocytes, have also been found in a diploid Chinese hamster cell line (CHEF/18) following exposure to 5-aza-CR or 5-aza-CdR (SAGER and KOVAK 1982). The analogue induces the expression of the emetine-sensitive phenotype in emetine-resistant Chinese hamster somatic cell hybrids (WORTON et al. 1983). This phenotypic change occurred in 15%–20% of Emt[R] revertants surviving 5-aza-CR treatment, a frequency of conversion higher than that obtained with any other system reported to date. In addition, 5-aza-CR induced the expression of thymidine kinase (TK) in TK-deficient Chinese hamster cells (HARRIS 1982).

The effects of these analogues on differentiation, therefore, appear to be quite general, and have been confirmed with mouse, human, and hamster cells. The results with the diploid Chinese hamster cells are particularly significant, in view of the fact that these cells have an apparently stable diploid chromosome complement.

7 X Chromosome Reactivation by 5-Azacytidine

One of the two X chromosomes is inactive in the somatic cells of normal mammalian females (LYON 1961). The pattern of inactivation remains constant for that cell and its descendants, and the mechanism by which the genes on the inactive X chromosome are rendered nonexpressible has been suggested to be mediated by DNA methylation (RIGGS 1975).

MOHANDAS et al. (1981) exposed a mouse-human somatic cell hybrid, deficient in hypoxanthine guanine phosphoribosyl transferase (HPRT) and containing a structurally normal inactive human X chromosome, to 5-aza-CR and tested for the reactivation of human X-linked HPRT gene. The frequency of HPRT-positive clones after 5-aza-CR treatment was 1000 times greater than that observed in untreated hybrid cells. GRAVES (1982) has reported similar results using a mouse-mouse somatic cell hybrid, and LESTER et al. (1982) has confirmed these studies using a different mouse-human hybrid cell line. In all these studies the expression of two other X-linked markers, phosphoglycerate kinase and glucose-6-phosphate dehydrogenase, was also detected, although to varying degrees.

In contrast to these results, WOLF and MIGEON (1982) were not able to induce the expression of genes located on inactive X chromosomes in clonal populations of human skin fibroblasts. The one revertant colony obtained was suggested to have arisen due to a mutational mechanism, although we have not found 5-aza-CR to have significant mutagenic activity in mammalian cells (LANDOLPH and JONES 1982). The failure of 5-aza-CR to induce expression of the HPRT locus in these cells may have been due to cellular and experimental conditions which were not sufficiently permissive for new gene expression.

More recently we have shown that the reactivation of the HPRT locus requires that the hybrid cells be treated with the analogue in the latter part of the S phase of the cell cycle, when the inactive X chromosome is replicating (JONES et al. 1982). The data also showed that two division cycles were required after analogue treatment for the maximal rate of enzyme expression, implying that symmetrical demethylation of specific sites might be necessary for gene activity.

VENOLIA et al. (1982) and LESTER et al. (1982) have found that the change induced by 5-aza-CR in the inactive X chromosome appears to be established at the level of the DNA. DNA extracted from cells treated with 5-aza-CR which are active in HPRT expression can function to transfer resistance to selective medium to nontreated recipient cells. In contrast, DNA extracted from hybrid cells which had not been treated with 5-aza-CR is inactive in the gene transfer procedure utilized.

Together, therefore, these data provide evidence that 5-aza-CR activates genes by direct DNA alteration, and strongly support the hypothesis that X inactivation in the hybrid cells is mediated by DNA methylation.

8 Induction of Viral Gene Expression by 5-Azacytidine

The induction of endogenous viral sequences by 5-aza-CR was first reported by ALTANEROVA in 1972. 5-Aza-CR and several mutagenic and carcinogenic agents induced the formation of infectious viral particles from a virogenic line of hamster cells previously transformed by the Schmidt-Rupin strain of Rous sarcoma virus.

Transient exposure of chicken cells containing the endogenous inactive chicken retrovirus evl to 5-aza-CR results in the transcriptional activation of the virus and hypomethylation at specific sites within the genome (GROUDINE et al. 1981). An interesting observation in these experiments was that although 5-aza-CR induced a clear undermethylation of the evl sequences, it had no effect on the methylation states of several globin genes located within the same cells. This result had the important ramification that some type of selective corrective mechanism led to the remethylation of the globin genes within the chicken cells.

5-Azacytidine induces the expression of Balb virus 1 and Balb virus 2 from mouse cells, and ecotropic endogenous virus from AKR 2B cells (NIWA and SUGAHARA 1981). The efficiency of induction in these cases was high, and comparable to that of 5-bromodeoxyuridine. The level of methylcytosine (5 mC) in newly synthesized DNA was substantially reduced by 5-aza-CR treatment, and there was an inverse relationship between the level of DNA modification and the frequency of virus expression in these cells.

BEN-SASSON and KLEIN (1981) have found that 5-aza-CR can activate the Epstein-Barr virus cycle in latently infected human lymphoid lines. More recent studies by CLOUGH et al. (1982) have shown that 5-aza-CR can induce the reexpression of the herpes simplex TK gene from mouse cells transformed with the viral information in an inactive state. The analogue also inhibited the decay of TK expression in the TK$^+$ transformants removed from selective conditions. Restriction analysis showed that the inactive gene was methylated, and became unmethylated at specific sites after 5-aza-CR treatment.

These experiments demonstrate that 5-aza-CR can induce the expression of viral information from hamster, mouse, chicken, and human cells, indicating that the drug may have use in probing the mechanisms by which viral gene expression is suppressed at the DNA level in animal cells. It is also important to note that the chicken and TK systems showed a clear correlation between the state of methylation of specific sites within the viral genome and gene expression, and that these patterns underwent heritable change following exposure to the nucleoside analogue.

9 Inhibition of DNA Methylation by 5-Azacytidine

Most experimental evidence supports the hypothesis that the induction of new gene expression by 5-aza-CR is linked to its ability to inhibit the methylation of newly synthesized DNA. Our early studies (JONES and TAYLOR 1980) showed that 5% substitution of cytosine residues by 5-aza-cytosine resulted in greater than 80% inhibition of DNA methylation. These results suggested that incorporated 5-azacytosine strongly inhibited the methylation of cytosine residues at unsubstituted methylation sites.

We have used the ability of 5-aza-CR to inhibit the methylation of newly replicated DNA to prepare defined hemimethylated substrates, characterized with respect to 5-azacytosine and 5mC contents (JONES and TAYLOR 1981). Substitution of 5-azacytosine for cytosine resulted in decreasing content of 5mC in the newly replicated strand, and therefore in increasing numbers of available hemimethylated sites. However, substitution of up to 5% of the cytosine residues in total DNA by 5-azacytosine resulted in almost complete deficiency of 5mC in the newly replicated strand (JONES and TAYLOR 1981). These DNAs were efficient acceptors of methyl groups in an in vitro reaction in the presence of a crude preparation of DNA methyltransferase from mouse spleen (JONES and TAYLOR 1981). The in vitro reaction reflected the specificity of the methyltransferase within the cell, in that methyl groups from S-adenosylmethionine (SAM) were applied specifically to cytosine residues located in the undermethylated strand. Increasing levels of 5-azacytosine in the DNA generated increasing numbers of hemimethylated sites, but fewer of these sites could be subsequently methylated in vitro as the 5-azacytosine content of the DNA increased (JONES et al. 1983). Thus, DNAs containing low levels of incorporated 5-azacytosine are efficient methyl acceptors in the test tube, whereas DNA extracted from cells treated with high levels of the analogue contain too much 5-azacytosine to be efficient substrates.

These defined, hemimethylated DNA substrates were used to further characterize the mechanism of action of DNA methyltransferase partially purified from mouse spleen, and of its inhibition by 5-azacytosine-containing DNA. The early studies by DRAHOVSKY and MORRIS (1971 a, b) suggested that the enzyme was highly processive, rather than distributive (KORNBERG 1981) in that DNA methyltransferase remained associated with the substrate after each catalytic event and "walked" along the DNA substrate, scanning the helix for available methylation sites. We have found that DNA methyltransferase binds nonspecifically to duplex or single-stranded DNA, irrespective of its state of methylation, and remains associated with that DNA even after the addition of hemimethylated duplex DNA to reaction mixtures (TAYLOR and JONES 1982 b). These results support the idea of a highly processive enzyme.

The presence of high levels of 5-azacytosine in duplex, hemimethylated DNA resulted in the formation of a tight-binding complex between enzyme and substrate in vitro which could not be dissociated by increasing salt concentrations (TAYLOR and JONES 1982 b). The formation of a tight-binding complex in vivo could result in the inhibition of cytosine methylation at sites downstream from the site of substitution of 5-azacytosine.

TANAKA et al. (1980) showed that treatment of Ehrlich's ascites cells with 5-aza-CR led to a rapid decrease in extractable DNA methyltransferase activity. CREUSOT et al. (1982) reported a similar finding with Friend erythroleukemia cells, and showed that the decrease in extractable enzyme activity was rapid and time-and-dose dependent. In addition, DNA prepared from E. coli K12 which had been treated with 5-aza-CR irreversibly inhibited the action of the bacterial DNA (cytosine-5) methylase in vitro (FRIEDMAN 1979, 1981). In contrast to the mammalian DNA methyltransferases, the bacterial modification enzyme does not act in a processive manner, but the mechanism of inhibition by 5-aza-CR appears to be similar for both types of enzymes.

We have also found that treatment of cultured mouse embryo cells with 5-aza-CR and other cytidine analogues modified in the 5 position results in a loss of extractable, active enzyme (TAYLOR and JONES 1982 b). This occurred at concentrations of the analogues that were active in inducing changes in phenotypic expression and inhibiting DNA methylation in these cells. Furthermore, the recovery of extractable, active enzyme from treated cells occurred slowly after treatment (TAYLOR and JONES 1982 b). This suggested that the inhibition of DNA methyltransferase might be irreversible, and that new enzyme synthesis might be required to overcome the block of DNA methylation.

We have therefore proposed that DNA methyltransferase scans the DNA for methylation sites and forms a tight-binding complex with 5-azacytosine incorporated into newly replicated DNA. It is not clear whether incorporation of the analogue into a methylation site is required for inhibition of cytosine methylation, but the formation of a tight-binding complex could deplete the cell of methyltransferase. Thus, newly generated methylation sites downstream from the site of incorporation of the fraudulent base might escape methylation during the replication cycle (TAYLOR and JONES 1982b). This would explain how substitution of 5% of cytosine residues by 5-azacytosine might inhibit DNA methylation by 80%.

10   Inhibition of Transfer RNA Methylation by 5-Azacytidine

Administration of 5-aza-CR and 5-fluorocytidine to mice leads to a rapid time-and-dose-dependent decrease in tRNA cytosine-5-methyltransferase activity in mouse liver (LU and RANDERATH 1979, 1980; LU et al. 1976). Treatment with these compounds led to the synthesis of tRNA specifically lacking in 5mC, and RNA synthesis was required for the induction of enzyme inhibition in vivo (LU and RANDERATH 1980). A small RNA species (4S–7S) was isolated, which specifically inhibited the action of tRNA methyltransferase in vitro (LU and RANDERATH 1980). Thus, incorporation of the fraudulent base into a low-molecular-weight RNA fraction (perhaps tRNA) might be responsible for the observed inhibition of tRNA methyltransferase.

The mechanism of inhibition of tRNA cytosine-5-methyltransferases by cytidine analogues modified in the 5 position appears to be analogoues to the inhibition of DNA methyltransferases by these compounds.

References

Altanerova V (1972) Virus production induced by various chemical carcinogens in a virogenic hamster cell line transformed by Rous sarcoma virus. J Natl Cancer Inst 49:1375–1380

Anderson EP (1973) Nucleoside and nucleotide kinases. In: Boyer PD (ed) The enzymes. Academic, New York, pp 49–96

Anderson EP, Brockman RW (1964) Feedback inhibition of uridine kinase by cytidine triphosphate and uridine triphosphate. Biochim Biophys Acta 91:380–386

Beisler JA, Abbasi MM, Driscoll JS (1976) Dihydro-5-azacytidine hydrochloride, a biologically active and chemically stable analog of 5-azacytidine. Cancer Treat Rep 60(11):1671–1674

Ben-Sasson SA, Klein G (1981) Activation of the Epstein-Barr virus genome by 5-aza-cytidine in latency-infected human lymphoid lines. Int J Cancer 28:131–135

Bodner AJ, Ting RC, Gallo RC (1981) Induction of differentiation of human promyelocytic leukemia cells (HL-60) by nucleosides and methotrexate. J Natl Cancer Inst 67:1025–1030

Caputto R (1962) Nucleotide kinases. In: Boyer PD, Lardy H, Myrback K (eds) The enzymes. Academic, New York, pp 133–138

Chabner BA, Johns DG, Coleman CN, Drake JC, Evans WH (1974) Purification and properties of cytidine deaminase from normal and leukemic granulocytes. J Clin Invest 53:922–931

Chou TC, Burchenal JM, Fox JJ, Watanabe KA, Chu CK, Philips FS (1979) Metabolism and effects of 5 (β-D-ribofuranosyl) isocytidine in P815 cells. Cancer Res 39:720–728

Cihak A (1978) Transformation of 5-aza-2'-deoxycytidine-^{3}H and its incorporation in different systems of rapidly proliferating cells. Eur J Cancer 14:117–124

Cihak A, Sorm F (1965) Biochemical effects and metabolic transformations of 5-azacytidine in *Escherichia coli*. Collect Czech Chem Commun 30:2091–2101

Clough DW, Kunkel LM, Davidson RL (1982) 5-Azacytidine-induced reactivation of a herpes simplex thymidine kinase gene. Science 216:70–73

Compere SJ, Palmiter RD (1981) DNA methylation controls the inducibility of the mouse metallothionein -1 gene in lymphoid cells. Cell 25:233–240

Constantinides PG, Jones PA, Gevers W (1977) Functional striated muscle cells from non-myoblast precursors following 5-azacytidine treatment. Nature (Lond) 267:364–366

Constantinides PG, Taylor SM, Jones PA (1978) Phenotypic conversion of cultured mouse embryo cells by azapyrimidine nucleosides. Dev Biol 66:57–71

Constantinides PG, Scott-Burden T, Gevers W (1980) Contrasting activity of two 5-aza-analogs of cytidine in the induction of muscle differentiation in cultured mouse embryo cells. Biochem Inter 1:24–31

Creusot F, Acs G, Christman J (1982) Inhibition of DNA methyltransferase and induction of Friend erythroleukemia cell differentiation by 5-aza-cytidine and 5-aza-2'-deoxycytidine. J Biol Chem 257:2041–2048

Diasio RB, Bennett JE, Myers CE (1978) Mode of action of 5-fluorocytosine. Biochem Pharmacol 27:703–707

Drahovsky D, Morris NR (1971a) Mechanism of action of rat liver DNA methylase. I. Interaction with double-stranded methyl-acceptor DNA. J Mol Biol 57:475–489

Drahovsky D, Morris NR (1971b) Mechanism of action of rat liver DNA methylase. II. Interaction with single-stranded methyl-acceptor DNA. J Mol Biol 61:343–356

Friedman S (1979) The effect of 5-azacytidine on *E. coli* DNA methylase. Biochem Biophys Res Commun 89:1328–1333

Friedman S (1981) The inhibition of DNA (cytosine-5) methylases by 5-azacytidine. The effect of azacytosine-containing DNA. Mol Pharmacol 19:314–320

Futterman B, Derr J, Biesler JA, Abbasi MM, Voytek P (1978) Studies on the cytostatic action, phosphorylation and deamination of 5-azacytidine and 5,6-dihydro-5-azacytidine in HeLa cells. Biochem Pharmacol 27:907–909

Glazer RI, Peale AL, Biesler JA, Abbasi MM (1980) The effect of 5-azacytidine and dihydro-5-azacytidine on nuclear ribosomal RNA and poly (A) RNA in L1210 cells in vitro. Mol Pharmacol 17:111–117

Graves JAM (1982) 5-Azacytidine induced re-expression of the alleles on the inactive X-chromosome in a *Mus musculus* × *M. cardi* cell line. Exptl Cell Res 141:99–105

Groudine M, Eisenman R, Weintraub H (1981) Chromatin structure of endogenous retroviral genes and activation by an inhibitor of DNA methylation. Nature (Lond) 292:311–317

Halle S (1968) 5-Azacytidine as a mutagen for arboviruses. J Virol 2:1228–1229

Harris M (1982) Induction of thymidine kinase in enzyme-deficient Chinese hamster cells. Cell 29:483–492

Jones PA, Taylor SM (1980) Cellular differentiation, cytidine analogs and DNA methylation. Cell 20:85–93

Jones PA, Taylor SM (1981) Hemimethylated duplex DNAs prepared from 5-azacytidine-treated cells. Nucleic Acids Res 9:2933–2947

Jones PA, Benedict WF, Baker MS, Mondal S, Rapp V, Heidelberger C (1976) Oncogenic transformation of C3H/10T$^1/_2$ clone 8 mouse embryo cells by halogenated pyrimidine nucleosides. Cancer Res 36:101–107

Jones PA, Taylor SM, Mohandas T, Shapiro LJ (1982) Cell cycle-specific reactivation of an inactive X-chromosome locus by 5-azadeoxycytidine. Proc Natl Acad Sci USA 79:1215–1219

Jones PA, Taylor SM, Wilson VL (1983) Inhibition of DNA methylation by 5-azacytidine and chemical carcinogens. In: Workshop on gene transfer and cancer. Raven, New York (in press)

Karon M, Benedict WF (1972) Chromatid breakage: differential effect of inhibitors of DNA synthesis during G2 phase. Science 178:62

Kornberg A (1981) DNA polymerases – a perspective. In: Boyer PD (ed) The enzymes. Academic, New York, pp 3–13

Landolph JR, Jones PA (1982) Mutagenicity of 5-azacytidine and related nucleosides in C3H/ $10T^1/_2$/CL8 and V79 cells. Cancer Res 42:817–823

Lee TT, Karon M (1976) Inhibition of protein synthesis in 5-azacytidine-treated HeLa cells. Biochem Pharmacol 25:1737–1742

Lee TT, Karon M, Momparler RL (1974) Kinetic studies on phosphorylation of 5-azacytidine with the purified uridine-cytidine kinase from calf thymus. Cancer Res 34:2482–2488

Lee TT, Karon M, Momparler RL (1975) Cellular phosphorylation of 1-β-D-arabinofuranosyl-cytosine and 5-azacytidine with intact fibrosarcoma and leukemic cells. Cancer Res 35:2506–2510

Lester SC, Korn NJ, DeMars R (1982) Derepression of genes on the human inactive X chromosome: evidence for differences in locus-specific rates of derepression and rates of transfer of active and inactive genes after DNA-mediated transformation. Somatic Cell Genet 8:265–284

Li LH, Olin EJ, Buskirk HH, Reineke LM (1970) Cytotoxicity and mode of action of 5-azacytidine on L1210 leukemia. Cancer Res 30:2760–2769

Liacouras AS, Anderson EP (1979) Uridine-cytidine kinase IV. Kinetics of the competition between 5-azacytidine and the two natural substrates. Mol Pharmacol 15:331–340

Lu L-JW, Randerath K (1979) Effects of 5-azacytidine on transfer RNA methyltransferases. Cancer Res 39:940–948

Lu, L-JW, Randerath K (1980) Mechanism of 5-azacytidine-induced transfer RNA cytosine-5-methyltransferase deficiency. Cancer Res 40:2701–2705

Lu L-JW, Chiang GH, Medina D, Randerath K (1976) Drug effects on nucleic acid modification. I. A special effect of 5-azacytidine on mammalian transfer RNA methylation in vivo. Biochem Biophys Res Commun 68:1094–1101

Lucas ZJ (1967) Pyrimidine nucleoside synthesis: regulatory control during transformation of lymphocytes in vitro. Science 156:1237–1240

Lyon MF (1961) Gene action in the X chromosome of the mouse (*Mus musculus* L.). Nature (Lond) 190:372–373

Maley F, Maley GF (1960) Nucleotide interconversions. II. Elevation of deoxycytidylate deaminase and thymidylate synthetase in regenerating rat liver. J Biol Chem 235:2968–2970

Mohandas T, Sparkes RS, Shapiro LJ (1981) Reactivation of an inactive human X chromosome: evidence for X inactivation by DNA methylation. Science 211:393–396

Momparler RL, Goodman J (1977) In vitro cytotoxic and biochemical effects of 5-aza-2'-deoxycytidine. Cancer Res 37:1636–1639

Momparler RL, Derse D (1979) Kinetics of phosphorylation of 5-aza-2'-deoxycytidine by deoxycytidine kinase. Biochem Pharmacol 28:1443–1444

Mondal S, Heidelberger C (1980) Inhibition of induced differentiation of $C3H/10T^1/_2$ clone 8 mouse embryo cells by tumor promoters. Cancer Res 40:334–338

Mondal S, Brankow DW, Heidelberger C (1976) Two-stage chemical oncogenesis in cultures of $C3H/10T^1/_2$ cells. Cancer Res 36:2254–2260

Nesnow S, Heidelberger C (1976) The effect of modifiers of microsomal enzymes on chemical oncogenesis in cultures of C3H mouse cell lines. Cancer Res 36:1801–1808

Niwa O, Sugahara T (1981) 5-Azacytidine induction of mouse endogenous type-C virus and suppression of DNA methylation. Proc Natl Acad Sci USA 78:6290–6294

Notari RE, DeYoung JL (1975) Kinetics and mechanisms of degradation of the antileukemic agent 5-azacytidine in aqueous solutions. J Pharmacol Sci 64:1148–1157

Piskala A, Sorm F (1964) Nucleic acid components and their analogs. LI. Synthesis of 1-glycosyl derivatives of 5-azauracil and 5-azacytosine. Collect Czech Chem Commun 29:2060–2076

Pithova P, Piskala A, Pitha J, Sorm F (1965) Nucleic acid components and their analogs. LXVI. Hydrolysis of 5-azacytidine and its connection with biological activity. Collect Czech Chem Commun 30:2801–2811

Pringle CR (1970) Genetic characteristics of conditional lethal mutants of vesicular stomatitis virus induced by 5-fluorouracil, 5-azacytidine and ethyl methane sulfonate. J Virol 5:559–567

Riggs AD (1975) X inactivation, differentiation and DNA methylation. Cytogenet Cell Genet 14:9–25

Rutter WJ, Pictet RL, Morris PW (1973) Toward molecular mechanisms of developmental processes. Annu Rev Biochem 42:601–646

Sager R, Kovac P (1982) Pre-adipocyte determination either by insulin or by 5-azacytidine. Proc Natl Acad Sci USA 79:480–484

Skold O (1960) Uridine kinase from Ehrlich ascites tumor: Purification and properties. J Biol Chem 235:3273–3279

Sorm F, Sormova Z, Raska K Jr, Jurovcik M (1966) Comparison of the metabolism and inhibitory effects of 5-azacytidine and 5-aza-2′-deoxycytidine in mammalian tissues. Rev Roum Biochim 3(1):139–147

Stoller RG, Coleman CN, Chang P, Hande KR, Chabner BA (1976) Biochemical pharmacology of cytidine analog metabolism in human leukemia cells. Bibl Haematol 43:531–533

Tanaka M, Hibasami H, Nagai J, Ikeda T (1980) Effect of 5-azacytidine on DNA methylation in Ehrlich's ascites tumor cells. Aust J Exp Biol Med Sci 58:391–396

Taylor SM, Jones PA (1979) Multiple new phenotypes induced in $10T^1/_2$ and 3T3 cells treated with 5-azacytidine. Cell 17:771–779

Taylor SM, Jones PA (1982a) Changes in phenotypic expression in embryonic and adult cells by 5-azacytidine. J Cell Physiol 111:187–194

Taylor SM, Jones PA (1982b) Mechanism of action of eukaryotic DNA methyltransferase: use of 5-azacytosine containing DNA. J Mol Biol 162:679–692

Venolia L, Gartler SM, Wassman ER, Ven P, Mohandas T, Shapiro LJ (1982) Transformation with DNA from 5-azacytidine-reactivated X chromosomes. Proc Natl Acad Sci USA 79:2352–2354

Vesely J, Cihak A (1977) Incorporation of a potent antileukemic agent, 5-aza-2′-deoxycytidine, into DNA of cells from leukemic mice. Cancer Res 37:3684–3689

Vesely J, Cihak A (1978) 5-Azacytidine: mechanism of action and biological effects in mammalian cells. Pharmac Ther A 2:813–840

Vesely J, Cihak A, Sorm F (1969) Biochemical mechanisms of drug resistance. IX. Metabolic alterations in leukemic mouse cells following 5-aza-2′-deoxycytidine. Collect Czech Chem Commun 34:901–909

Viegas-Pequignot E, Dutrillaux B (1976) Segmentation of human chromosomes induced by 5-ACR (5-azacytidine). Hum Genet 34:247–254

Voytek P, Beisler JA, Abbasi MM, Wolpert-De Filippes MK (1977) Comparative studies on the cytostatic action and metabolism of 5-azacytidine and 5,6-dihydro-5-azacytidine. Cancer Res 37:1956–1961

Wilson VL, Jones PA, Momparler RL (1983) Inhibition of DNA methylation in L1210 leukemic cells by 5-aza-2′-deoxycytidine. A possible mechanism of chemotherapeutic action. Cancer Res 43:3493–3496

Wolf SF, Migeon BR (1982) Studies of X chromosome DNA methylation in normal human cells. Nature (Lond) 295:667–671

Worton RG, Grant S, Duff C (1983) Gene inactivation and reactivation at the *emt* locus in Chinese hamster cells. In: Workshop on gene transfer and cancer. Raven, New York (in press)

Zadrazil S, Fucik V, Bartl P, Sormova Z, Sorm F (1965) The structure of DNA from *E. coli* cultured in the presence of 5-azacytidine. Biochim Biophys Acta 108:701–703

Methylation of the Genes for 18S, 28S, and 5S Ribosomal RNA

A.P. BIRD

1 Introduction

In animals, the modified base 5-methylcytosine (5mC) is found predominantly in the sequence CpG (WYATT 1951; VANYUSHIN et al. 1970; DOSCOCIL and SORM 1962; GRIPPO et al. 1968). Not all CpGs, however, are methylated. With the introduction of restriction endonucleases as probes for CpG methylation, it became possible to determine the relative locations of a subset of methylated and unmethylated CpGs along the DNA (GAUTIER et al. 1977; BIRD and SOUTHERN 1978; BIRD 1978; WAALWIJK and FLAVELL 1978). The information from this kind of experiment is most precise when a specific sequence is studied, and it has proved to be particularly interesting when the sequence concerned is a gene whose transcription is regulated. This chapter describes the methylation pattern of genes coding for 18S, 28S and 5S ribosomal RNA. These genes exhibit the exciting – and the perplexing – aspects of DNA methylation in relation to gene expression.

Ribosomes contain three kinds of RNA molecule. The two larger molecules, 18S rRNA and 28S rRNA, are synthesized as a single precursor molecule, which is processed to give the mature rRNA (e.g. WELLAUER and DAWID 1974). Throughout higher animals and plants the genes coding for 18S and 28S RNA are tandemly repeated, head to tail, at one or more loci in the genome. The repeated unit of ribosomal DNA (rDNA) comprises

MRC Mammalian Genome Unit, King's Buildings, West Mains Road, Edinburgh EH9 3JT Scotland, United Kingdom

Current Topics in Microbiology and Immunology, Vol. 108
© Springer-Verlag Berlin · Heidelberg 1984

a nontranscribed spacer plus a region coding for the rRNA precursor (BIRN-STIEL et al. 1968; BROWN and WEBER 1968; DAWID et al. 1970; WENSINK and BROWN 1971). Surprisingly, the smallest ribosomal RNA, 5S RNA, is encoded by a separate family of genes, though it is required in amounts equimolar with 18S and 28S rRNA. The 5S genes, like 18S and 28S genes, comprise a family of tandem repeats in which coding regions alternate with nontranscribed spacer (BROWN et al. 1971).

The genes for 18S and 28S rRNA offer several experimental advantages for studying DNA methylation. First, the genes are large and contain many potentially methylatable CpGs. Thus there are many sites in and around the coding region whose methylation state can be studied. Second, the sequence of the coding region is highly conserved in evolution (BIRNSTIEL et al. 1971; GERBI 1976; SINCLAIR and BROWN 1971). As a result, a hybridization probe that is prepared from one species can be used to reliably detect the rDNA from a variety of animal, plant and fungal species. Wide-ranging comparative studies are therefore easily accomplished. Third, since ribosomal genes have been well studied in many species, experiments on their patterns of methylation can be related to known aspects of their organization and function in the cell.

Detailed analysis of 5S DNA methylation is presently confined to a single species, *Xenopus laevis* (BROWN et al. 1971; FEDOROFF and BROWN 1978; MILLER et al. 1978). It is significant, however, because the position of every 5mC moiety has been localized by direct sequencing of the genomic DNA (see Sect. 5.4).

2 Interspecies Variation in Methylation of rDNA

For a long time it has been known that the level of 5mC in genomic DNA varies widely among living organisms (WYATT 1951; VANYUSHIN et al. 1970; CHARGAFF et al. 1952). In insects 5mC is not detectable at a sensitivity of 1 in 1000 C residues (A. Razin, personal communication), while in vertebrates about 5% of C residues are methylated (data collected by SHAPIRO 1968). Still higher levels occur in green plants, where as much as 30% of C is commonly methylated (SHAPIRO 1968).

Interspecies variability in methylation is also evident at the level of a single gene family. The rDNA of many invertebrates is not detectably methylated at those CpGs that are testable with restriction endonucleases (RAE and STEELE 1979; BIRD and TAGGART 1980). In contrast, chromosomal rDNA of some vertebrates (e.g. amphibia and fish, BIRD and TAGGART 1980) and of several green plants (G. Clark and A. Bird, unpublished observations) is heavily methylated. Between these extremes, many mammals exhibit a combination of methylated and unmethylated rDNA (see Sect. 3).

Studies of total DNA and of rDNA in a variety of systems suggest that methylated and unmethylated sequence domains coexist in the eukaryotic genome. Compartmentalization of this kind is most evident in genomes where the two fractions are of comparable size. For example, when sea

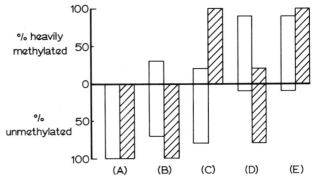

Fig. 1. The distribution of total DNA (*open bars*) and rDNA (*shaded bars*) between methylated and unmethylated fractions of the genome. Each of the patterns (*A*) to (*E*) represents one combination that has been observed. At one extreme (*A*) are many insects in which 100% of rDNA and total DNA is unmethylated (within the sensitivity of the experiments). At the other extreme are some vertebrates and green plants in which most genomic DNA and all rDNA is in the heavily methylated compartment. In cases where the DNA is divided between heavily methylated and unmethylated compartments, the values are not precise. For example, the fraction of methylated rDNA in mammals (pattern *D*) ranges from 0% to 30%, and the value indicated is 20%. The species exhibiting each pattern are itemized in Table 1

urchin DNA is digested with CpG enzymes (i.e. enzymes whose recognition sequence contains a CpG, and which are blocked by methylation of the C) and fractionated on an agarose gel, the fragments are separated into a low-molecular-weight, unmethylated fraction, and a high-molecular-weight fraction that is heavily methylated (BIRD et al. 1979). Each fraction comprises long sequence domains of between 15 kb and over 50 kb in length, which either lack detectable 5mC, or are heavily methylated at CpG. Similar experiments with vertebrate DNA do not detect an obvious unmethylated sequence compartment, as most of the DNA is poorly digested by CpG enzymes. When rDNA is examined, however, it is found to be predominantely unmethylated in a wide range of vertebrates (seven mammals, two birds and a reptile; BIRD and TAGGART 1980). The results show that unmethylated domains exist in vertebrates and suggest that methylation compartments of the kind in many invertebrates may be widespread. This suggestion is borne out by recent experiments on total genomic DNA (D. Cooper and A. Bird, unpublished observations). If vertebrate DNA is terminally labelled after digestion with CpG enzymes, an unmethylated fraction of DNA can be detected by autoradiography. Terminal labelling gives a distribution of fragment numbers (since each fragment is labelled at its ends, regardless of its length), whereas normal ethidium bromide staining gives the DNA distribution by weight. As a result, small fragments which bind very low levels of ethidium bromide can be easily visualized by end labelling. The results suggest that methylation compartments are widespread in nature, and that the variable levels of 5mC in different organisms arise through changes in the relative size of methylated and unmethylated compartments. Figure 1 summarizes the observed distributions of total genomic DNA and rDNA, and Table 1 details the species assigned to each kind of distribution.

Table 1. Distribution of total DNA and rDNA between methylated and unmethylated compartments in a wide range of organisms

Methylation pattern[a]	Phylum	Species	Common name	Tissue	Reference[b]
(A)	Arthropods	*Drosophila melanogaster*	Fruit fly	Adult	1
		Drosophila virilis	Fruit fly	Adult Embryo	2
		Psithyrus sp.	Bee	Adult	2
		Musca domestica	House fly	Adult	2
		Sarcophaga bullata	Flesh fly	Adult	2
		Cancer pagurus	Edible crab	Mixed	1
	Ascomycetes	*Saccharomyces cereviciae*	Yeast	Whole	4
(B)	Protists	*Physarum polycephalum*	Slime mould	Whole	8
	Coelenterates	*Metridium senile*	Sea anenome	Whole	1
	Molluscs	*Mytilus edulis*	Common mussel	Testis	1
	Echinoderms	*Echinus esculentus*	Sea urchin	Egg Sperm Embryo Intestine Tubefoot	3
		Paracentrotus lividus	Sea urchin	Sperm	1
		Strongylocentrotus purpuratus	Sea urchin	Sperm	1
		Psammechinus miliaris	Sea urchin	Sperm	1
		Asterias rubens	Starfish	Sperm Intestine	1 1
		Thyone fusus	Sea cucumber	Sperm	1
		Ophiopholis	Brittle star	Sperm	1
		Ophiothrix fragilis	Brittle star	Sperm	1
	Chordates	*Ciona intestinalis*	Sea squirt	Mixed	1
(C)	Angiosperms	*Brassica alba*	White mustard	Seedlings	4
(D)	Chordates (vertebrate)	*Elaphe radiate*	Indian krait (snake)	Liver	1
		Gallus domesticus	Chicken	Liver	1
		Coturnix coturnix	Quail	Liver	1
		Homo sapiens	Man	Placenta	1
		Mus musculus	Mouse	Liver Brain Testis Embryo	5

Table 1 (continued)

Methyl-ation pattern[a]	Phylum	Species	Common name	Tissue	Refer-ence[b]
		Rattus norvegicus	Rat	Liver Embryo Brain Jejunal Epithelium Sperm	6
		Oryctolagus cuniculus	Rabbit	Liver	1
		Ovis aries	Sheep	Liver	1
		Sus domesticus	Pig	Liver Sperm	1
		Bos taurus	Calf	Thymus	1
(E)	Vertebrates	*Salmo trutta*	Trout	Testis	1
		Xenopus laevis	Clawed frog	Liver Brain Sperm Embryo Tadpole Blood	7
		Triturus cristatus	Crested newt	Blood	1
	Angiosperms	*Triticum aestinum*	Wheat	Seedlings	4
		Secale cereale	Rye	Seedlings	4
		Pisum sativum	Pea	Seedlings	4
		Nicotiana tabacum	Tobacco	Seedlings	4
		Phaseolus aureus	Mung bean	Seedlings	4

[a] Patterns (A) to (E) are those diagrammed in Fig. 1. All patterns were established using restriction endonucleases as probes for DNA methylation

[b] References: 1. BIRD and TAGGART 1980 – 2. RAE and STEELE 1979 – 3. BIRD et al. 1979 – 4. CLARK CG, BIRD AP, unpublished results – 5. BIRD et al. 1981a – 6. KUNNATH and LOCKER 1982 – 7. BIRD et al. 1981b – 8. REILLY et al. 1980

It is apparent that nearly all possible permutations have been observed. No current model for the function of DNA methylation fully accounts for this variability.

3 Methylated and Unmethylated Ribosomal Genes in Mammals

In the invertebrates, rDNA is part of the unmethylated compartment, while in many amphibia, fish and green plants it is part of the methylated compartment. Surprisingly, in many other vertebrates it appears that the family of rDNA repeats is divided between both compartments. A well-studied example is the mouse (BIRD et al. 1981a). Most mouse rDNA is digested to the same extent by *Hpa*II and *Msp*I, suggesting that it is unmethylated at these sites. However, a minority of DNA sequences that hybridize to the rDNA probe are very poorly digested by *Hpa*II and other CpG enzymes. That the undigested sequences are indeed rDNA can be demonstrated by cleaving them with a range of restriction enzymes that are not affected by CpG methylation. The fragments produced are characteristic of mouse rDNA. The amount of rDNA in each fraction (i.e. heavily methylated or unmethylated) does not change perceptibly when different tissues, including sperm and whole embryos, are examined, and only slight variation is detectable between individuals of the same inbred strain. Between inbred strains, however, there is considerable variability. For example, 30% of rDNA in CBA mice is methylated, compared with about 5% in C57 black mice.

Methylated and unmethylated fractions of rDNA have also been observed in the rat (KUNNATH and LOCKER 1982). The fraction resistant to cleavage by CpG enzymes was about 30% of the total rDNA. The proportion of methylated rDNA appeared lower in 14-day embryos than in later embryos or in adult tissues, though the effect was not quantified. Changes in the methylated fraction of rDNA were clearly detected, however, when embryonic and adult tissues were compared. Repeat units that were heavily methylated in the embryo had lost some methylation at discrete sites in adult tissues. The location of the undermethylated regions has not yet been reported.

What is it that determines whether a sequence is methylated or unmethylated? Work on rDNA suggests that it is not the nucleotide sequence immediately surrounding a CpG, because unmethylated and methylated ribosomal genes exhibit no obvious sequence differences, as judged by restriction mapping (BIRD et al. 1981a). In addition, it is evident that methylation, or lack of it, occurs in sequence domains, suggesting that the unit of methylation is not the individual CpG, but a domain that contains many CpGs. It is possible that the signal to methylate is not a sequence, but a chromosomal location. Alternatively, the timing of replication at the locus may be important. Unfortunately, we know little about the chromosomal location or time of replication of methylated and unmethylated ribosomal genes in mice and rats. It would be interesting, for example, to determine whether methylated genes are clustered at one of the multiple nucleolar organizers present in both these species, or whether they are interspersed with unmethylated repeats at each nucleolar locus.

4 Unmethylated rDNA in *Xenopus* Germ Cells

As discussed above, chromosomal rDNA in *Xenopus* cells is heavily methylated at CpG. During gametogenesis, however, extrachromosomal rDNA copies are generated (MILLER 1966; BROWN and DAWID 1968; GALL 1968; EVANS and BIRNSTIEL 1968) and these amplified genes are not methylated.

The phenomenon was first detected as a buoyant density difference between amplified rDNA (1.729 g/ml) and somatic rDNA (1.723 g/ml) in caesium chloride gradients (BROWN and DAWID 1968). Base composition analysis subsequently showed that the only detectable difference between the two rDNAs was the presence of 5mC in somatic rDNA (DAWID et al. 1970). No 5mC was detectable in amplified rDNA. In this study amplified rDNA was purified from oocytes, as the level of amplification is several thousand fold in these cells. Lower levels of rDNA amplification (5- to 40-fold) occur in premeiotic germ cells of both sexes (GALL 1969). Buoyant density experiments on this rDNA indicate that it, too, lacks methylation of C (KALT and GALL 1974; BIRD 1978).

5 rDNA Methylation and Transcription

5.1 Evidence for Differential Expression of Methylated and Unmethylated rDNA Fractions in Mammals

Most ribosomal RNA genes in the mouse are unmethylated, but there is a significant fraction of methylated genes. In view of correlations in other systems between lack of methylation and transcription (RAZIN and RIGGS 1980) it is important to know whether methylated and unmethylated fractions of rDNA are differentially transcribed. In practice, this experiment cannot be approached directly because of the difficulty in distinguishing transcripts from methylated and unmethylated genes. Expressing genes are known, however, to be preferentially sensitive to DNase I in isolated nuclei, due to a change in chromatin structure associated with transcription (WEINTRAUB and GROUDINE 1976). Isolated nuclei from livers of Balb/c mice were therefore treated with DNase I and the sensitivity of methylated and unmethylated genes was compared (BIRD et al. 1981a). The results showed that unmethylated rDNA is preferentially sensitive to DNase I. Though indirect, this experiment suggests that unmethylated rDNA is transcriptionally more active than methylated rDNA. Thus, the correlation established for several endogenous and viral genes also appears to hold true for mouse rDNA.

A similar correlation is found in a rat hepatoma cell line in which the rDNA content per cell is ten times higher than that in diploid cells (TANTRAVAHI et al. 1981). Most of the extra genes are inactive by the criterion of silver staining, and analysis of the rDNA indicates that most is heavily methylated at HpaII and HhaI sites. As in the mouse, heavily methylated rDNA does not appear to be transcriptionally active.

5.2 Transcription of Heavily Methylated rDNA in Xenopus

The chromosomal rDNA of Xenopus laevis is heavily methylated at CpG. Closer examination of rDNA from blood or liver cells reveals, however, that methylation within the repeat unit is not uniform (BIRD and SOUTHERN

1978). In somatic cells a region of the nontranscribed spacer is deficient in methylation, as judged by its susceptibility to the CpG enzymes *Hpa*II and *Ava*I (Fig. 2; BIRD et al. 1981 b). The undermethylated region is coincident with two clusters containing a 60-bp repeated sequence, occasionally interspersed with an unrelated 21-bp sequence (BOSELEY et al. 1979; SOLLNER-WEBB and REEDER 1979; MOSS et al. 1980). Each copy of the 60-bp sequence contains one *Hpa*II site and one *Ava*I site, and it is these sites that are undermethylated. The remainder of *Hpa*II sites (over 150 per repeat unit) and *Ava*I sites (over 100 per repeat unit) are 98%–99% methylated. The situation is different in sperm rDNA. Here the nontranscribed spacer is refractory to *Hpa*II and *Ava*I, indicating that the region is fully methylated (BIRD et al. 1981 b). The transition from a fully methylated spacer to an undermethylated spacer occurs during the first day of embryonic development (Fig. 2). Thus it correlates in time with the onset of rDNA transcription in the embryo (BROWN and LITTNA 1964).

Methylation in a related species, *Xenopus borealis*, shows striking parallels with *X. laevis*, but also some important differences. Undermethylated regions are present in the nontranscribed spacer, and their location corresponds closely with that in *X. laevis* rDNA (MACLEOD and BIRD 1982). Furthermore, the sequence surrounding the hypomethylated sites in *X. borealis* rDNA is very similar to the 60-bp subrepeat that is undermethylated in *X. laevis* rDNA (LA VOLPE et al. 1982). Thus, the location and nucleotide sequence of the undermethylated regions is conserved between these species. In spite of this similarity there is a major developmental difference between *X. borealis* and *X. laevis* with respect to the undermethylated sites. In *X. laevis* the sites are fully methylated in sperm and undermethylated in somatic tissues, but in *X. borealis* they are undermethylated in both sperm and somatic tissue (MACLEOD and BIRD 1982). Thus, the striking temporal correlation between loss of methylation and appearance of rRNA transcription that is seen in *X. laevis* does not take place in *X. borealis* embryos. Since there is no evidence for differential regulation of embryonic rRNA synthesis between the species (BROWN and LITTNA 1964; HONJO and REEDER 1973), it follows that a change in the pattern of rDNA methylation as observed in *X. laevis* embryos is not an obligatory feature of ribosomal gene regulation.

5.3 Absence of a Link Between DNase I Hypersensitivity and Undermethylation

The absence of an intimate relationship between spacer undermethylation and rDNA transcription is also demonstrated by experiments with hybrids of *X. borealis* and *X. laevis*. Transcription of *X. borealis* ribosomal genes is suppressed in hybrid tadpoles (HONJO and REEDER 1973), and, correspondingly, the genes are not in a DNase I-hypersensitive configuration (MACLEOD and BIRD 1982). Nevertheless, undermethylated sites are present in the nontranscribed spacer of inactive *X. borealis* rDNA. We conclude that spacer undermethylation is not a sufficient condition for transcription. Studies of

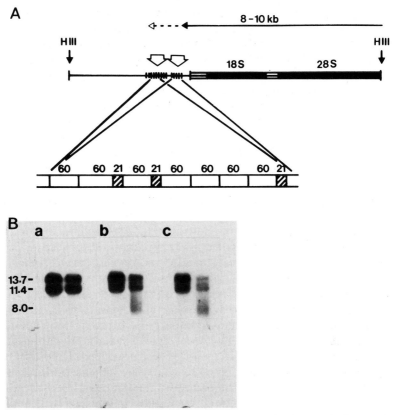

Fig. 2 A, B. Appearance during development of an undermethylated region in rDNA of *Xenopus laevis*. **A** The rDNA repeat unit flanked by sites for *Hin*dIII (*HIII*). The nontranscribed spacer (*thin line*) contains undermethylated sites for *Hpa*II and *Ava*I. Double digestion with *Hin*dIII and *Hpa*II or *Ava*I gives a broad band of 8–10 kb (*top*) due to cleavage at the repeated sequences. Each repeat unit contains over 150 sites for *Hpa*II and over 100 sites for *Ava*I, but sites outside the repeated sequence are immune to digestion by virtue of CpG methylation. *Below* is a blowup of the internally repeated regions showing 60-bp repeats, occasionally interrupted by 21-bp repeats. **B** Appearance of the undermethylated sites during development. Paired digests are of *X. laevis* sperm DNA (*a*), blastula DNA (*b*), and heartbeat-stage DNA (*c*). Each sample was digested with *Hin*dIII (*left*) and *Hin*dIII plus *Hpa*II (*right*), and rDNA was detected by blotting (SOUTHERN 1975) and hybridization to a labelled rDNA probe. Sperm rDNA is not digested by *Hpa*II, but during blastula a diffuse band at 8–10 kb appears due to *Hpa*II cleavage in the spacer. By heartbeat, most repeats have lost methylation in the nontranscribed spacer. (Adapted from MACLEOD and BIRD 1982)

human β-like globin genes also demonstrate that undermethylation does not automatically lead to transcription (VAN DER PLOEG and FLAVELL 1980) since placenta and KB cells, which do not express globin genes, are undermethylated at the locus to a greater extent than expressing cells. Similarly, in the sea urchin there is no detectable methylation of the histone genes, either in early embryos which are actively transcribing the bulk of histone genes, or in sperm and gastrulae, where most histone genes are not being transcribed (BIRD et al. 1979). Finally, *Drosophila*, in which methylated CpG is not detectable (RAE and STEELE 1979; BIRD and TAGGART 1980), is nevertheless able to convert genes from active to inactive chromatin configuration (WU et al. 1979). More comparable to the rDNA data are the results of a study of the chicken α2(I) collagen gene (MCKEON et al. 1982). In this case, an unmethylated domain surrounds and includes the 5′ end of the gene, while most of the coding sequence is a methylated domain. This methylation pattern holds in all the tissues tested, including sperm and other nonexpressing cells. DNase-I hypersensitivity correlates with transcription, but not with the presence of the unmethylated 5′ regions. Thus, as in the case of *X. borealis* rDNA, localized undermethylation is not closely tied to gene expression.

These results demonstrate that localized or generalized undermethylation of a gene is not sufficient to ensure its transcription. The many correlations between lack of methylation and transcription, could be explained, however, if undermethylation were a necessary precondition for transcription. Experiments to test this possibility are in progress in several laboratories, and have yielded preliminary results that are consistent with the idea that methylation inhibits transcription (VARDIMON et al. 1982; STEIN et al. 1982). Equivalent experiments with *Xenopus* rDNA are also currently underway. Specifically, the question is to compare the transcription capacity of sperm and blood rDNA from *Xenopus laevis*. If spacer methylation inhibits transcription, then sperm rDNA will not be transcribed. Ribosomal genes offer the advantage over other systems that they can be purified in the native methylation state by physical methods (WALLACE and BIRNSTIEL 1966). Most genes can only be purified by cloning, and thereby lose their native methylation patterns.

5.4 Transcription and Methylation of Xenopus laevis 5S Genes

Another gene family that can be purified using physical methods alone is the 5S gene family of *Xenopus laevis* (BROWN et al. 1971). The majority of 5S genes in *Xenopus* are reserved for use in the growing oocyte (FORD and SOUTHERN 1973), and it is exclusively oocyte-type genes that have been purified in this way. Oocyte-type 5S DNA comprises a rather homogeneous family of tandem repeat units, each about 450 bp in length (BROWN et al. 1971). The purified fraction is sufficiently homogeneous to allow direct sequencing, and since the MAXAM and GILBERT (1980) technique detects 5mC, it is possible to locate all sites of C methylation in the repeat unit. Using this method, FEDOROFF and BROWN (1978) and MILLER et al. (1978) showed that all CpG sequences within the repeat unit are heavily methylated. Since the genes were purified from blood cells, where they are not transcribed, it was of interest to determine whether they can be transcribed. To test this, 5S DNA from blood cells was injected into oocytes (BROWN and GURDON 1977). The result showed that purified 5S DNA, though heavily methylated, is as good a template for RNA synthesis as cloned 5S DNA, which lacks CpG methylation. On the face of it, this result contradicts correlations which indicate that heavily methylated genes are transcriptionally inert, but there are questions which remain. For example, is it significant that the molecules were linear rather than circular? Is the oocyte a good model system for assaying the effects of methylation? Are genes that are transcribed by RNA polymerase III atypical? In spite of these questions this system has provided one of the most direct tests to date of the effects of methylation on transcription. The significance of the results will have to be assessed in the light of experience with other genes in this and other transcription systems.

References

Bird AP (1978) A study of early events in ribosomal gene amplification. Cold Spring Harbor Symp Quant Biol 42:1179–1183

Bird AP (1978) Use of restriction enzymes to study eukaryotic DNA methylation. II. The symmetry of methylated sites supports semiconservative copying of the methylation pattern. J Mol Biol 118:49–60

Bird AP, Southern EM (1978) Use of restriction enzymes to study eukaryotic DNA methylation. I. The methylation pattern in ribosomal DNA from *Xenopus*. J Mol Biol 118:27–47

Bird AP, Taggart MH (1980) Variable patterns of total DNA and rDNA methylation in animals. Nucleic Acids Res 8:1485–1497

Bird AP, Taggart MH, Smith BA (1979) Methylated and unmethylated DNA compartments in the sea urchin genome. Cell 17:889–901

Bird AP, Taggart MH, Gehring C (1981a) Methylated and unmethylated ribosomal RNA genes in the mouse. J Mol Biol 152:1–17

Bird AP, Taggart MH, Macleod D (1981b) Loss of rDNA methylation accompanies the onset of ribosomal gene activity in early development of *X. laevis*. Cell 26:281–390

Birnstiel M, Speirs J, Purdom I, Jones K, Leoning UE (1968) Properties and composition of the isolated ribosomal DNA satellite of *Xenopus laevis*. Nature 219:454–463

Boseley PG, Moss T, Machler M, Portmann R, Birnstiel ML (1979) Sequence organisation of the spacer DNA in a ribosomal gene unit of *Xenopus laevis*. Cell 17:19–31

Brown DD, Littna E (1964) RNA synthesised during the development of *Xenopus laevis*, the South African clawed toad. J Mol Biol 8:669–687

Brown DD, Dawid IB (1968) Specific gene amplification in oocytes. Science 160:272–280

Brown DD, Weber CS (1968) Gene linkage by RNA-DNA hybridisation. II. Arrangement of the redundant gene sequences for 28S and 18S rRNA. J Mol Biol 34:681–698

Brown DD, Gurdon J (1977) High-fidelity transcription of 5S DNA injected into *Xenopus* oocytes. Proc Natl Acad Sci USA 74:2064–2068

Brown DD, Wensink PC, Jordan E (1971) Purification and some characteristics of 5S DNA from *Xenopus laevis*. Proc Natl Acad Sci USA 68:3175–3179

Chargaff E, Lipshitz R, Green C (1952) Composition of the deoxypentose nucleic acids of 4 genera of sea urchin. J Biol Chem 195:155–166

Dawid IB, Brown DD, Reeder RH (1970) Composition and structure of chromosomal and amplified ribosomal DNAs of *Xenopus laevis*. J Mol Biol 51:341–360

Doscocil J, Sorm F (1962) Distribution of 5-methylcytosine in pyrimidine sequences of deoxyribonucleic acids. Biochim Biophys Acta 55:953–959

Evans D, Birnstiel M (1968) Localisation of amplified ribosomal DNA in the oocyte of *Xenopus laevis*. Biochim Biophys Acta 166:2740–276

Fedoroff NV, Brown DD (1978) The nucleotide sequence of oocytes 5S DNA in *Xenopus laevis*. I. The AT-rich spacer. Cell 13:701–716

Ford PJ, Southern EM (1973) Different sequences for 5S RNA in kidney cells and ovaries of *Xenopus laevis*. Nature (New Biol) 241:7–10

Gall JG (1968) The genes for ribosomal RNA during oogenesis. Genetics [Suppl] 61:121–131

Gautier F, Bunemann H, Grotjahn L (1977) Analysis of calf thymus satellite DNA: evidence for specific methylation of cytosine in CG sequences. Eur J Biochem 80:175–183

Gerbi S (1976) Fine structure of ribosomal RNA. I. Conservation of homologous regions within ribosomal RNA of eukaryotes. J Mol Biol 106:791–816

Grippo P, Iaccarino M, Parisi E, Scarano E (1968) Methylation of DNA in developing sea urchin embryos. J Mol Biol 36:196–208

Honjo T, Reeder RH (1973) Preferential transcription of *Xenopus laevis* ribosomal RNA in interspecies hybrids between *Xenopus laevis* and *Xenopus mulleri*, J Mol Biol 80:217–228

Kalt MR, Gall JG (1974) Observations on early germ cell development and premeiotic ribosomal DNA amplification in *Xenopus laevis*. J Cell Biol 62:460–472

Kunnath L, Locker J (1982) Variable methylation of the ribosomal RNA genes of the rat. Nucleic Acids Res 10:3877–3892

La Volpe A, Taggart MH, Macleod D, Bird AP (1982) Coupled demethylation of sites in a conserved sequence of *Xenopus* ribosomal DNA. Cold Spring Harbor Symp Quant Biol 47:585–592

Macleod D, Bird A (1982) DNase I sensitivity and methylation of active versus inactive rRNA genes in *Xenopus* species hybrids. Cell 29:211–218

Maxam AM, Gilbert W (1980) Sequencing end-labelled DNA with base-specific chemical cleavages. Methods Enzymol 65:499–560

McKeon C, Ohkubo H, Pastan I, de Crombrugghe B (1982) Unusual methylation pattern of the $\alpha 2(1)$ collagen gene. Cell 29:203–210

Miller OL (1966) Structure and composition of peripheral nucleoli of salamander oocytes. Natl Cancer Inst Monogr 23:53–66

Miller OL, Cartwright EM, Brownlee GG, Fedoroff NV, Brown DD (1978) The nucleotide sequence of oocytes 5S DNA in *Xenopus laevis*. II. The GC-rich region. Cell 13:717–725

Moss T, Boseley PG, Birnstiel ML (1980) More spacer sequences of *X. laevis* rDNA. Nucleic Acids Res 8:467–485

Rae PMM, Steele RE (1979) Absence of cytosine methylation at CCGG and GCGC sites in the rDNA coding regions and intervening sequences of *Drosophila* and the rDNA of other higher insects. Nucleic Acids Res 6:2987–2995

Razin A, Riggs AD (1980) DNA methylation and gene function. Science 210:604–610

Reilly JG, Braun R, Thomas CA (1980) Methylation in *Physarum* DNA. Febs Lett 116:181–184

Shapiro HS (1968) Distribution of purines and pyrimidines in deoxyribose-nucleic acids. In: Sober HA (ed) Handbook of biochemistry. The Chemical Rubber Co., Cleveland, pp H39–H48

Sinclair JH, Brown DD (1971) Retention of common nucleotide sequences in ribosomal deoxyribonucleic acid of eukaryotes and some of their characteristics. Biochemistry 10:2761–2769

Sollner-Webb B, Reeder RH (1979) The nucleotide sequence of the initiation and termination sites for ribosomal RNA transcription in *X. laevis*. Cell 18:485–499

Southern EM (1975) Detection of specific sequences among DNA fragments separated by gel electrophoresis. J Mol Biol 98:503–517

Stein R, Razin A, Cedar H (1982) In vitro methylation of the hamster adenine phosphoribosyl transferase gene inhibits its expression in mouse L cells. Proc Natl Acad Sci USA 79:3418–3422

Tantravahi U, Ramareddy VG, Erlanger BF, Miller OJ (1981) Amplified ribosomal RNA genes in a rat hepatoma cell line are enriched in 5-methylcytosine. Proc Natl Acad Sci USA 78:489–493

Van der Ploeg LHT, Flavell RA (1980) DNA methylation in the human $\gamma\delta\beta$-globin locus in erythroid and nonerythroid tissues. Cell 19:947–958

Vanyushin BF, Tkacheva SG, Belozersky AN (1970) Rare bases in animal DNA. Nature 225:948–951

Vardimon L, Kressmann A, Cedar H, Maechler M, Doerfler W (1982) Expression of a cloned adenovirus gene is inhibited by in vitro methylation. Proc Natl Acad Sci USA 79:1073–1077

Waalwijk C, Flavell RA (1978) MspI, an isoschizomer of *Hpa*II which cleaves both unmethylated and methylated *Hpa*II sites. Nucleic Acids Res 5:3231–3236

Wallace H, Birnstiel ML (1966) Ribosomal cistrons and the nucleolar organizer. Biochem Biophys Acta 114:296–310

Weintraub H, Groudine M (1976) Chromosomal subunits in active genes have an altered conformation. Science 193:848–856

Wellauer PK, Dawid IB (1974) Secondary structure maps of ribosomal RNA and DNA. I. Processing of *Xenopus laevis* ribosomal RNA and structure of single-stranded ribosomal DNA. J Mol Biol 89:379–395

Wensink PC, Brown DD (1971) Denaturation map of the ribosomal DNA of *X. laevis*. J Mol Biol 60:235–248

Wu C, Wong YC, Elgin SCR (1979) The chromatin structure of specific genes. II. Disruption of chromatin structure during gene activity. Cell 16:807–814

Wyatt GR (1951) The purine and pyrimidine composition of deoxypentose nucleic acids. Biochem J 48:584–590

Eukaryotic DNA Methylase – Properties and Action on Native DNA and Chromatin

R.L.P. Adams, T. Davis, J. Fulton, D. Kirk, M. Qureshi, and R.H. Burdon

1 Introduction

DNA methylase (*S*-adenosyl-L-methionine DNA (cytosine-5) methyltrans-ferase: EC 2.1.1.37) is an enzyme which catalyses the transfer of methyl groups from *S*-adenosylmethionine (AdoMet) to certain cytosine residues in DNA (Borek and Srinivasan 1966). The product of the reaction is DNA containing 5 methylcytosine. In eukaryotes most of the cytosines methylated in vivo are found in the dinucleotide CG but not all cytosines in this dinucleotide are methylated (Sinsheimer 1955; Doskocil and Sorm 1962; Singer et al. 1979). Methylcytosine is also found in other dinucleo-tides, e.g. mCC, but the origin and significance of these forms have not been investigated.

Because only a proportion of CG dinucleotides are methylated, and because, to some extent, the pattern of methylation of a particular DNA sequence is passed on from one generation to the next (Wigler et al. 1981), a maintenance methylase has been postulated which, following DNA repli-cation, will add methyl groups to the daughter strand opposite methylated CG dinucleotides.

```
mCG     parent
 GC     daughter
 ↑
Methylase acts here
```

Department of Biochemistry, University of Glasgow, Glasgow G12 8QQ, Scotland, United Kingdom

Current Topics in Microbiology and Immunology, Vol. 108
© Springer-Verlag Berlin · Heidelberg 1984

Indeed, most methylation of DNA occurs shortly after replication (KAPPLER 1970; BURDON and ADAMS 1969) and the preferred substrate for mouse DNA methylase is duplex DNA in which one strand contains and the other lacks methylcytosines, i.e. hemimethylated DNA (ADAMS et al. 1979a; GRUENBAUM et al. 1982). However, this may not be a universal feature of animal DNA methylases, as the rat liver enzyme shows no strong preference for hemimethylated DNA over unmethylated duplex (SIMON 1982). The pattern of methylation is not identical in all cells of an organism, but appears to depend in part on the transcriptional status of a particular DNA sequence. Thus, during chick development there is undermethylation of those globin genes which are being expressed (WEINTRAUB et al. 1981). In order for a change in the methylation pattern to occur, specific removal or addition of methyl groups to certain cytosine residues is necessary. Removal would require the activity of a "demethylase" enzyme, or simply DNA replication in the absence of methylase activity (BURDON and ADAMS 1980). Addition, on the other hand, would require the action of methylase on hitherto unmethylated sequences of DNA. Whilst such methylation de novo might seem necessary only during embryogenesis, it is puzzling that this type of activity is exhibited by DNA methylases isolated from mature somatic tissues, and no evidence for the presence of a second distinct DNA methylase in the cells of a higher eukaryote has yet appeared.

We have been looking at the activity of a DNA methylase which can be extracted from nuclei of mouse Krebs II ascites cells with buffer solutions containing 0.2–0.4 M NaCl (TURNBULL and ADAMS 1976; ADAMS et al. 1979a). We chose these cells as we found the nuclear extracts to have the highest specific activity of DNA methylase, and the cells are readily available in large amounts (ADAMS et al. 1974). Other workers have extracted DNA methylase activity from rat liver, spleen, Novikoff hepatoma cells (SHEID et al. 1968; KALOUSEK and MORRIS 1969; MORRIS and PIH 1971; SIMON et al. 1978; SNEIDER et al. 1975), and HeLa cells (ROY and WEISSBACH 1975). In addition, DNA methylase activity has been extracted from *Xenopus laevis* cells (ADAMS et al. 1981), chick embryo (L.A. McLelland and R.L.P. Adams, unpublished), and *Chlamydomonas* (SANO and SAGER 1980).

Apart from the last report cited, there is no evidence for multiple species of DNA methylase, and repeated extraction of nuclei with increased salt concentrations fails to remove any further activity. However, we find that when all the soluble DNA methylase activity has been removed from Krebs II ascites cell nuclei these remains behind activity which sediments with the "nuclear matrix" in 2 M NaCl (QURESHI et al. 1982). This bound form of DNA methylase, however, appears not to be a different enzyme but, by virtue of possible association with the DNA replicating machinery of the matrix it may perform a methyl maintenance function in vivo. Conversely, because of this we would suggest that the soluble DNA methylase, acting on regions of DNA unmethylated in both strands, could well bring about methylation de novo.

It is also likely that a major function of the soluble enzyme in the nucleus of a somatic cell is to act on those hemimethylated sites which fail to be methylated immediately following replication. The presence of such hemimethylated sites is suggested by (a) the delay between DNA synthesis and its complete methylation (ADAMS 1971; Table 3); (b) the continuation of DNA methylation when synthesis is blocked by drugs (BURDON and ADAMS 1969); (c) the ability of isolated nuclei to add methyl groups to the recently synthesized strand of endogenous DNA (ADAMS and HOGARTH 1973); and (d) the ability of DNA methylase to add methyl groups to homologous DNA in vitro, an ability which is greatest when that DNA comes from cells starved for methionine (TURNBULL and ADAMS 1976). Such hemimethylated sites in eukaryotic DNA may arise as a result of interference with methylase function by proteins associated with the DNA, or as a consequence of postreplicative repair of DNA.

2 Purification and Assay

The soluble DNA methylase referred to has been partially purified by the research groups listed earlier, using phosphocellulose chromatography and ammonium sulphate fractionation, gel filtration, and separation on DEAE. Estimates of the native molecular weight range from 120000 to 180000. DNA methylase activity is assayed by measuring the transfer of methyl groups from S-adenosyl-[^{3}H-Me]-L-methionine to DNA. The acceptor DNA is purified after the incubation by phenol extraction and ethanol precipitation. RNA is degraded by alkali treatment, and acid soluble radioactivity removed by washing with 5% trichloracetic acid and ethanol (pronase and ribonuclease treatments are also sometimes included). The final DNA can be hydrolyzed with formic acid to the bases which, on separation – on an Aminex A6 column, for instance – show the radioactivity to be almost completely in 5-methylcytosine (ADAMS and BURDON 1983). A small amount of radioactivity in thymine probably arises due to deamination of 5-methylcytosine during acid hydrolysis (FORD et al. 1980). The specific activity of the enzyme depends on the DNA substrate used, and the highest reported so far is 4680 pmol methyl groups transferred per mg protein per h for the rat liver enzyme, using denatured DNA from *Micrococcus luteus* (SIMON et al. 1978). However, much higher specific activities may be obtained if hemimethylated DNA is used as substrate (GRUENBAUM et al. 1982).

3 Action of DNA Methylase on Unmethylated Native DNA

3.1 Procession of DNA Methylase

It is clear from several studies using bacterial or insect DNA as substrate that DNA methylase can add methyl groups to native DNA which is essentially unmethylated, i.e. DNA which contains CG dinucleotides which are unmethylated in both strands (DRAHOVSKY and MORRIS 1971; ADAMS et al. 1979a, b; SIMON et al. 1978). The reaction is slow, however, in that methyl groups continue to be added over many hours, and it is extremely difficult to methylate all CG dinucleotides.

From early work with rat liver DNA methylase, DRAHOVSKY and MORRIS (1971) postulated a mode of action for the enzyme on native DNA. They suggested that the enzyme binds to DNA, possibly at a transiently denatured site, and then travels along the DNA, methylating cytosines in a salt-stimulated reaction. They reached this conclusion from experiments showing that while addition of 0.2 M NaCl to the assay at zero time inhibited the reaction, addition at 3 min had a delayed effect, the extent of delay depending on the length of the DNA substrate. They calculated the rate of enzyme movement to be about 2.5 nucleotides/s. This was based on unidirectional movement of the enzyme and a complete (or nearly complete) modification of all suitable cytosines.

Similar results were obtained with mouse DNA methylase when native unmethylated DNA from mosquito cells was used as substrate, but we could find little evidence of complex formation with methylated native DNA from calf thymus, mouse L929, or *Xenopus* cells (ADAMS et al. 1979a). We did, in addition, obtain evidence for the importance of transiently denatured sites, in that, with increase in incubation temperature, the relative efficiency of methylation of "native" DNA versus single-stranded DNA increased sixfold.

To investigate this problem further we have used the circular replicative form of φX174 DNA as substrate, for our partially purified methylase (which is free of nuclease as judged by its inability to degrade supercoiled DNA). In a reaction in which the amount of methylase is limiting, two alternative mechanisms of reaction may be envisaged:

1. The enzyme binds to a particular site on the DNA, methylates it, and immediately dissociates from the DNA.
2. The enzyme binds to DNA and travels along the molecule methylating available sites. The initial binding site may or may not be a site of methylation. This is the mechanism favoured by DRAHOVSKY and MORRIS (1971), as mentioned above.

In the second alternative, after a reaction of the enzyme with an excess of defined DNA molecules, e.g. duplex cyclic φX174 DNA, some DNA molecules would be fully methylated and others would be unmethylated. This result contrasts to that expected if the first alternative were correct, whereby most DNA molecules would be partially methylated.

Following incubation in vitro with our DNA methylase, the φ174 DNA is recovered and incubated with the restriction endonuclease *Hpa*II which cleaves at the unmethylated sequence CCGG (MANN and SMITH 1978). Normally, the DNA of φX174 RF would be cleaved five times to give five fragments (SANGER et al. 1977) but methylation of the central cytosine in the CCGG sequence would block the activity of *Hpa*II nuclease (MANN and SMITH 1978). Thus, if a DNA methylase were to methylate all CpG sites including the five *Hpa*II sites, the φX174 DNA would become resistant to *Hpa*II.

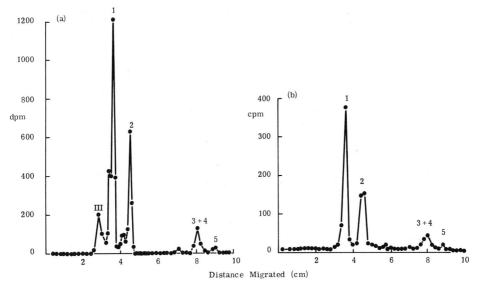

Fig. 1 a, b. Results of methylation. **a** *Hpa*II cleavage of methylated DNA. 3.5 µg φX174 RF DNA was incubated at 37° for 3 h with [³H] AdoMet and DNA methylase (6 µg protein). The reaction was terminated by heating to 60° for 10 min. Ten units *Hpa*II was added with the appropriate buffer and the mixture incubated for a further 2 h at 37°. A mixture of sodium dodecylsulfate, glycerol and bromphenol blue was added and the sample applied to a 1.5% agarose tube gel. After electrophoresis the gel was stained with ethidium bromide and washed several times with 5% trichloracetic acid before slicing into 1- or 2-mm slices. Radioactivity was assayed as previously described (Adams et al. 1979b). **b** Methylation of *Hpa*II fragments. Four µg φX174 RF DNA was incubated with 5 units *Hpa*II at 37° for 90 min; [³H] AdoMet, EDTA (15 mM) and DNA methylase (11 µg protein) were added and the incubation continued at 37°. Samples (40% of total) were taken at 30 min and 3 h and electrophoresed as described for **a**. The graph shows the results for the 30 min incubation

Figure 1 a shows the result of methylation of φX174 RF DNA by Krebs II ascites cell DNA methylase. The incubation with methylase was for 3 h, and from the results of subsequent HpaII digestion of the φX174 DNA the distribution of methyl groups is close to what would be expected if a random selection of φX174 DNA cytosines were methylated (Table 1), though the smaller *Hpa*II fragments are somewhat undermethylated. Thus, most of the φX174 DNA is still largely susceptible to *Hpa*II, despite the activity of the mouse DNA methylase.

However, about 7% of the methyl groups are in linear (form III) DNA (and some are in bands running ahead of fragment 1 and between fragments 1 and 2), showing that a small proportion of φX174 DNA molecules are heavily methylated and resistant to *Hpa*II endonuclease at four of the five sites.

In the reaction shown in Fig. 1 a, 1.6 methyl groups are added, on average, per φX174 DNA molecule. As there are 267 CpGs in single-stranded φX174 DNA, this could mean either that 0.3% of the duplex DNA molecules are fully methylated (and the remainder unmethylated) or that each

Table 1. Methylation of ϕX174 RF DNA *Hpa*II fragments

*Hpa*II fragment		1	2	3	4	5	Total
Base pairs (bp)		2745	1690	374	348	218	5375
				722			
CpGs		266	168	78		22	534
CpG/bp × 100		9.7	9.9	10.8		10.1	9.9
dpm	expt. 1a	1860	1050	255		50	3185
	expt. 1b	2150	1375	350		65	3940
dpm/bp	expt. 1a	0.68	0.62	0.32		0.23	0.59
	expt. 1b	0.78	0.81	0.48		0.30	0.73

DNA molecule received only one (or two) methyl groups, the enzyme leaving the DNA after each addition. These are the two extreme possibilities and intermediate explanations appear more probable.

Figure 1b shows the result when the substrate ϕX174 DNA was cleaved with *Hpa*II prior to incubation with DNA methylase. All fragments accept methyl groups, but again the smaller fragments are somewhat undermethylated (Table 1). This undermethylation may reflect an artefactual loss of small fragments, but, as it is present in both experiments (Figs. 1a and 1b), it is therefore not the result of preferential methylation of particular *Hpa*II sites.

When the in vitro methylation reaction was continued for up to 30 h it became clear (Fig. 2a) that a significant fraction of the substrate ϕX174 DNA molecules became partially resistant to *Hpa*II, as faint bands of linear ϕX174 DNA (5375 bp) and of a 3110-bp fragment could be seen even after prolonged digestion with *Hpa*II. A fluorograph of this gel (Fig. 2b) shows these two bands to be disproportionately heavily labelled with methyl groups, and reveals other methylated partial-digestion products of 2230 bp and 2000 bp. It is also of considerable interest that, although form-II ϕX174 DNA predominates in the incubation, the larger proportion of the methyl groups is in form I DNA prior to *Hpa*II digestion (Columns 3, 5 and 7 of Fig. 2).

The fact that no subset of fully methylated, *Hpa*II-resistant DNA molecules containing all the added methyl groups is produced on methylating an excess of circular DNA with our mouse DNA methylase appears to rule out the simple idea of DRAHOVSKY and MORRIS (1971), and in similar experiments with a rat liver methylase, SIMON (1982) could not obtain evidence for a processive mechanism. However, the production of a subset of partially resistant molecules lends some support to a partly processive mechanism or – alternatively – a cooperative binding of enzyme molecules to DNA. It is possible that 30 h is insufficient time for an enzyme molecule to travel around the cyclic ϕX174 DNA, but DRAHOVSKY and MORRIS (1971) calculated that the rate of movement is about 2.5 bp/s, which would give ample time for five cycles even in a 3-h incubation. Moreover, the rate

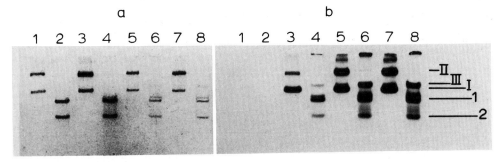

Fig. 2 a, b. Prolonged methylation of φX RF DNA. Ten μg φX174 RF DNA was incubated at 37° with [³H] AdoMet and DNA methylase (50 μg protein). Triplicate samples were removed after 6 h, 24 h and 30 h, and more enzyme and Ado Met added after 6 h and 24 h. One of the samples was used to assay total incorporation which came to 8.1, 29.8 and 40.9 methyl groups per φX174 RF DNA molecule after 6 h, 24 h and 30 h respectively. A second sample was incubated at 37° for 2 h with 5 units *Hpa*II, while the third sample acted as control. **a** Photograph of the samples electrophoresed on a slab gel. **b** Fluorogram of the same gel. Columns 2, 4, 6 and 8 show the *Hpa*II-treated samples and columns 1, 3, 5 and 7 the controls, following incubation with DNA methylase for 0, 6, 24 and 30 h respectively

of DNA synthesis in vivo is about 50 bases/s and any processing methylase would be expected to keep pace.

However, DRAHOVSKY and MORRIS (1971) also indicated that their enzyme adds only three methyl groups while travelling along a piece of DNA with a molecular weight of 1.8×10^6, i.e. less than 2% of CpGs methylated, and other data indicate that prolonged incubation times (30 h) are normally required to maximize the number of methyl groups added to a given DNA by mammalian enzymes (TURNBULL and ADAMS 1976; SIMON et al. 1978). If only occasional methyl groups are added the enzyme would have to complete many cycles of φX174 RF DNA before methylation is complete. However, the enzyme may well dissociate from DNA on encountering a site already methylated. In vivo, this would have the advantage that the enzyme is not involved in fruitless searches for potential sites in fully methylated DNA, and it may explain our difficulty in forming a salt-resistant complex with already methylated duplex DNAs.

3.2 Methylation of Supercoiled DNA

As was noted earlier, some evidence was obtained that form-I supercoiled DNA may be a better substrate for the mouse DNA methylase than form-II open circular DNA. Indeed, the presence of methyl groups in form-II DNA may arise by nicking of already methylated form-I DNA molecules during prolonged incubation. Brief treatment of form-I DNA with a single-stranded nuclease produces a single nick in the DNA, and prolonged treatment leads to a second nick opposite the first to yield linear DNA (BARTOK and DENHARDT 1976; LILLEY 1980).

150   R.L.P. Adams et al.

Table 2. Effect of single-stranded nuclease on acceptor activity of duplex φX174 DNA

Time of nuclease digestion (min)	Methyl groups added (dpm/h)
0	695
3	110
60	131

φX174 RF DNA (10 μg) was treated at 37° with 0.5 units *Neurospora crassa* nuclease in 75 mM Tris/HCl pH 8.0; 7.5 mM MgCl₂. Samples were removed after 0 min, 3 min or 1 h. The reaction was terminated by addition of EDTA (14 mM). Samples analysed by gel electrophoresis showed most of the DNA to be present as nicked circles after 3 min digestion, and as linear DNA with a little form-II DNA after 1 h digestion. DNA methylase and [³H]-*S*-adenosylmethionine was added to the DNA, which was incubated at 37° for 1 h

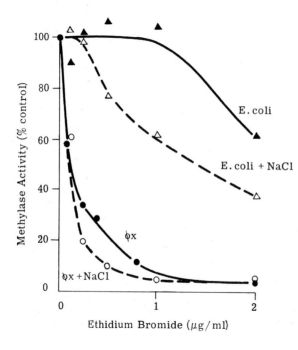

Fig. 3. Effect of ethidium bromide on DNA methylation. 1.1 μg φX174 RF DNA or native *E. coli* DNA was incubated for 1 h with DNA methylase (5 μg protein) in the presence of the indicated concentration of ethidium bromide; 0.1 *M* NaCl was present in those samples indicated

Table 2 gives the results of using as DNA substrate φX174 DNA which is (a) largely form I, (b) largely form II or (c) a mixture of form II and III. It is clear that form-I DNA is the best substrate.

Figure 3 shows the effect of ethidium bromide on the methylation reaction. Addition of ethidium bromide to supercoiled DNA leads first to an uncoiling as the dye intercalates between the bases. At a concentration of 2 μg/ml, ethidium bromide will convert φX174 form-I DNA into relaxed circles (DENHARDT and KATO 1973). However, the first few intercalations would be expected to remove the single-stranded regions from form-I molecules. Methylation of φX174 DNA is completely abolished with 1 μg/ml ethidium bromide, while methylation of *E. coli* DNA is only slightly affected.

The experiments reported here support the mode of action of DNA methylase where the enzyme binds to a single-stranded region in duplex DNA and moves at least some distance along the molecule. It appears that the enzyme can also bind less efficiently at nicks or ends of duplex DNA. The methylation of totally unmethylated DNA is very inefficient, and the enzyme may require several passages before methylation is in any way complete. In vivo the effect of the presence of chromosomal proteins may exert a controlling influence on which sites are methylated.

3.3 Methylation of SV40 DNA and SV40 Minichromosomes

Although we previously reported a failure to methylate SV40 DNA, we have now shown that form-I supercoiled SV40 DNA will accept methyl groups, albeit at a rate less than 20% that of form-I ϕX174 DNA. Gel electrophoresis shows that form-II SV40 DNA accepts very few methyl groups, although methylated form-II DNA arises on prolonged incubation, probably by nicking form-I DNA. Cleavage of methylated form-I SV40 DNA with *Hin*dIII gives six fragments, all of which are methylated in a 24-h reaction.

The probable reason for the poor acceptor activity of SV40 DNA relative to ϕX174 RF DNA may be that the former is deficient in CpG dinucleotides, having only 54 per molecule as compared with the 534 in ϕX174 RF DNA (REDDY et al. 1978).

Although SV40 DNA is capable of accepting methyl groups in vitro, the SV40 minichromosome does not have an exposed single-stranded region (HERMAN et al. 1979), and so may not be able to bind DNA methylase. (Single-stranded regions are present, however, in replicating SV40 minichromosomes.) We have been able to add a very small number of methyl groups to SV40 minichromosomes by incubating nuclei from SV40-infected BSCl cells with high-specific-activity [^{3}H-Me]-AdoMet and mouse DNA methylase. Analyses of SV40 form-I DNA purified from the incubation by SDS-sucrose gradient sedimentation or by Hirt extraction and agarose gel electrophoresis indicate that about 4% of the methyl groups incorporated by isolated nuclei were in supercoiled viral DNA.

Another possible explanation for the lack of methylation of SV40 DNA in vivo is that, following infection, there is an inhibition of DNA methylase activity. In contrast, we have found that, especially when cells in stationary culture are infected with SV40, there is a tenfold rise in the methylase activity recovered in nuclear extracts. This is similar to the effect of SV40 infection on activity of the enzymes involved in DNA synthesis.

It appears likely, then, that there is a lack of methylation of SV40 DNA because it initially escapes the attention of a maintenance methylase, as SV40 replication is not associated with the nuclear matrix. Subsequently, the lack of a suitable single-stranded entry point on the mature minichromosome, coupled with the inhibitory action of chromosomal proteins, precludes action by the soluble DNA methylase.

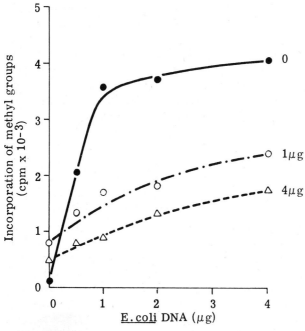

Fig. 4. Inhibition of methylation of *E. coli* DNA by DNA containing 5-azacytosine. Increasing amounts of native *E. coli* DNA were incubated with [³H] AdoMet and 24 μg mouse ascites methylase for 1 h in the presence of 0, 1 or 4 μg DNA, isolated from mouse L929 cells grown for 24 h in the presence of 10 μ*M* azadeoxycytidine

3.4 Affinity of DNA Methylase for Substrate DNA

As DNA methylase has not been obtained in pure form, no two preparations are identical. Therefore, it is particularly difficult to compare K_m values for DNA in light of the proposed mechanism of action on the DNA. Thus, the amount of DNA required to saturate the enzyme depends not only on the concentration of the DNA but also on the amount of enzyme present in the incubation. Using one enzyme preparation we have shown, for example, that in order to saturate a given amount of enzyme the concentration of unmethylated duplex DNA from mosquito is ten times lower than of methylated DNA from calf thymus are required to saturate a given amount of enzyme (ADAMS et al. 1979a). This implies that the enzyme may interact with unmethylated CG dinucleotides, and GRUENBAUM et al. (1981) have shown that the rate of reaction is proportional to the C + G content.

A comparison of DNA from control mouse L929 cells with DNA from cells grown for one generation in the presence of 1 μ*M* or 10 μ*M* azadeoxy-cytidine shows that the concentration required to achieve half maximal velocity falls with increasing substitution of the analogue, and it is possible that the enzyme binds extremely firmly to azaCG dinucleotides. Furthermore, the presence of azacytosine-substituted DNA in the methylase assay inhibits incorporation of methyl groups into native *E. coli* DNA (Fig. 4).

Table 3. In vivo methylation of chromosomal DNA; extent of methylation $\left(\dfrac{mC}{C+mC}\times100\right)^a$

Incubation conditions	Total DNA	Nucleosomal core DNA	Linker DNA[b]
Log cells – 55-min label	2.48 ± 0.53	3.90 ± 0.28	1.06
Log cells – 2-day label	4.00 ± 0.17	5.04 ± 0.11	2.96
Stat. cells – 2-day label ($\pm$2-day chase)	4.09 ± 0.44	5.06 ± 0.33	3.12
Log cells – labelled for 2 days, chased and subcultured	3.73 ± 0.22	4.91 ± 0.17	2.54

[a] The data are the mean $\pm$ standard deviation from 2–5 determinations
[b] The figure for the level of methylation of linker DNA is calculated assuming the 50% of DNA digested by the nuclease is all linker DNA

Mouse L929 cells were incubated with [6-^{3}H] uridine for the indicated times, nuclei prepared and a part digested with micrococcal nuclease (750 units/ml) for 30 min. Agarose gels were used to confirm that the product of digestion contained DNA less than 150 bp in length. DNA was isolated and digested with formic acid to the bases, which were separated on a column of Aminex A6 (ADAMS et al. 1979b)

This could explain the inhibitory effect of analogue substitution on methylation of adjacent CG dinucleotides (JONES and TAYLOR 1981; ADAMS et al. 1982), as well as the disappearance of methylase activity from nuclear extracts of cells treated with azadeoxycytidine and the absence of an associated decrease in the amount of enzyme in the matrix-bound form (TANAKA et al. 1980; CREUSOT et al. 1982; QURESHI et al. 1982b).

4 Methylation of Chromatin

The duplex DNA on which a soluble DNA methylase will act in a eukaryotic cell is complexed with histones and other proteins. These proteins may interfere with methylation in a specific, controlled way, or in a non specific way.

Incubation of chromatin from mosquito cells (which contains very little 5-methylcytosine; ADAMS et al. 1979b) with mouse DNA methylase and AdoMet leads to a preferential methylation of internucleosomal (linker) DNA rather than DNA associated with nucleosomal cores (T. Davis and R.L.P. Adams, manuscript in preparation). If a similar interference with methylation occurs in vivo, those regions of DNA on the leading side of the replication fork which are constantly associated with nucleosomes (SEIDMAN et al. 1979) may well become methylated only slowly.

However, the results presented in Table 3 indicate that complete methylation of both linker and core DNA is delayed in vivo. The in vitro rate of methylation of chromatin is only 10% that of naked DNA, implying a considerable inhibitory action of chromosomal proteins on the methylation of linker DNA (T. Davis and R.L.P. Adams, unpublished observations). Such an effect may contribute significantly to the undermethylation of those regions of chromatin undergoing frequent transcription, and hence most covered with associated proteins, such as RNA polymerase.

5 Conclusion

A considerable amount of work is still needed to completely purify and characterize eukaryotic DNA methylases. Despite the low specificity required, no evidence for multiple species of DNA methylases in higher eukaryotes is as yet available, and the distribution of methylated pyrimidine tracts found in vivo can be reproduced in vitro by the partially purified mouse DNA methylase (BROWNE et al. 1977).

The large enzyme from mouse ascites cells shows a marked preference for hemimethylated DNA, and binds more strongly to DNA containing a higher proportion of unmethylated CG dinucleotides, though binding must initially occur at a single-stranded region. The smaller rat liver enzyme shows little preference for hemimethylated DNA, and evidence for a processive mechanism is still controversial.

It should be remembered that, in vivo, the enzyme will always interact with DNA complexed with protein, and that, as well as serving a maintenance role, it must be able to carry out de novo methylation if DNA modification is to play a role in gene expression.

Acknowledgments. The authors would like to thank the MRC and the CRC for financial support, and Professor SMELLIE for encouragement over a number of years.

References

Adams RLP (1971) The relationship between synthesis and methylation of DNA in mouse fibroblasts. Biochim Biophys Acta 254:205–212

Adams RLP, Hogarth C (1973) DNA methylation in isolated nuclei: old and new DNA is methylated. Biochim Biophys Acta 331:214–220

Adams RLP, Burdon RH (1982) Eukaryotic DNA methylation. CRC Crit Revs Biochem 13:349–384

Adams RLP, Burdon RH (1983) DNA methylases. In: Hnilica L (ed) Enzymes of nucleic acid synthesis and processing, vol 2. 119–144. CRC Press, Cleveland

Adams RLP, Turnbull J, Smillie EJ, Burdon RH (1974) DNA methylation in nuclei and studies using a purified DNA methylase from ascites cells. In Postsynthetic modification of macromolecules. FEBS Symposium 34:39–48

Adams RLP, McKay EL, Craig LM, Burdon RH (1979a) Mouse DNA methylase: methylation of native DNA Biochim Biophys Acta 561:345–357

Adams RLP, McKay EL, Craig LM, Burdon RH (1979b) Methylation of mosquito DNA. Biochim Biophys Acta 563:72–81

Adams RLP, Burdon RH, Gibb S, McKay EL (1981) DNA methylase during *Xenopus laevis* development. Biochim Biophys Acta 655:329–334

Adams RLP, Fulton J, Kirk D (1982) The effect of 5-azadeoxycytidine on cell growth and DNA methylation. Biochim Biophys Acta 697:286–294

Bartok K, Denhardt DT (1976) Site of cleavage of superhelical φX174 RF DNA by the single-strand-specific *N. crassa* endonuclease. J Biol Chem 251:530–535

Borek E, Srinivasan PR (1966) The methylation of nucleic acids. Annu Rev Biochem 35:275–298

Browne MJ, Turnbull JF, McKay EL, Adams RLP, Burdon RH (1977) The sequence specificity of a mammalian DNA methylase. Nucleic Acids Res 4:1039–1045

Burdon RH, Adams RLP (1969) The in vivo methylation of DNA in mouse fibroblasts. Biochim Biophys Acta 174:322–329

Burdon RH, Adams RLP (1980) Eukaryotic DNA methylation. Trends Biochem Sci 5:294–297

Creusot F, Acs G, Christman JF (1982) Inhibition of DNA methyl transferase and induction of Friend erythroleukemia cell differentiation by 5-azacytidine and 5-azadeoxycytidine. J Biol Chem 257:2041–2048

Denhardt DT, Kato AC (1973) Comparison of the effect of UV radiation and ethidium bromide intercalation on the conformation of superhelical φX174 RF DNA. J Mol Biol 77:479–494

Doskocil J, Sorm F (1962) Distribution of 5 methylcytosine in pyrimidine sequences of DNA. Biochim Biophys Acta 55:953–959

Drahovsky D, Morris NR (1971) Mechanism of action of rat liver DNA methylase I: interaction with double stranded methyl-accepted DNA. J Mol Biol 57:475–489

Ford JP, Coca-Prados M, Hsu M-T (1980) Enzymatic analysis of 5-methylcytosine content in eukaryotic DNA. J Biol Chem 255:7544–7547

Gruenbaum Y, Stein R, Cedar H, Razin A (1981) Methylation of CpG sequences in eukaryotic DNA. FEBS Lett 124:67–71

Gruenbaum Y, Cedar J, Razin A (1982) Substrate and sequence specificity of a eukaryotic DNA methylase. Nature 295:620–622

Herman TM, De Pamphilis ML, Wassarman PM (1979) Structure of chromatin at DNA replication forks. Biochemistry 18:4563–4571

Jones PA, Taylor SM (1981) Hemimethylated duplex DNAs prepared from 5-azacytidine-treated cells. Nucleic Acids Res 9:2933–2947

Kalousek F, Morris NR (1969) The purification and properties of DNA methylase from rat spleen. J Biol Chem 244:1157–1163

Kappler J (1970) The kinetics of DNA methylation in cultures of a mouse adrenal cell line. J Cell Physiol 75:21–32

Lilley DMJ (1980) The inverted repeat as a recognizable structural feature in supercoiled DNA molecules. Proc Natl Acad Sci USA 77:6468–6472

Mann MB, Smith HD (1978) Specificity of Hpa II and Hae III DNA methylases. Nucleic Acids Res 4:4211–4221

Morris NR, Pih KD (1971) The preparation of soluble DNA methylase from normal and regenerating rat liver. Cancer Res 31:433–440

Qureshi M, Adams RLP, Burdon RH (1982) Soluble and bound forms of DNA methylase in mouse cell nuclei. Trans Biochem Soc 10:455–456

Reddy BV, Thimmappaya B, Dhar R, Subramanian KN, Zain BS, Pan J, Ghosh PK, Celma ML, Weissman SM (1978) The genome of simian virus 40. Science 200:494–502

Roy PH, Weissbach A (1975) DNA methylase from HeLa cell nuclei. Nucleic Acids Res 2:1669–1684

Sanger F, Air CM, Barrell BG, Brown NL, Coulson AR, Fiddes JC, Hutchison CA, Slocombe PM, Smith M (1977) Nucleotide sequence of bacteriophage φX174 DNA. Nature 265:687–695

Sano H, Sager R (1980) DNA methyltransferase from the eukaryote *Chlamydomonas reinhardi*. Eur J Biochem 105:476–480

Seidman MM, Levine AJ, Weintraub H (1979) The asymmetric segregation of parental nucleosomes during chromosome replication. Cell 18:439–449

Sheid B, Srinivasan PR, Borek E (1968) DNA methylase of mammalian tissues. Biochemistry 7:280–285

Simon D, Grunert F, von Acken U, Doring HP, Kroger H (1978) DNA methylase from regenerating rat liver: purification and characterisation. Nucleic Acids Res 5:2153–2167

Singer J, Robert-Ems J, Riggs AD (1979) Methylation of mouse liver DNA studied by means of the restriction enzymes *Msp*I and *Hpa*II. Science 203:1019–1021

Sinsheimer RL (1955) The action of pancreatic DNase, II isomeric dinucleotides. J Biol Chem 215:579–583

Sneider TW, Teague WM, Rogachevsky LM (1975) S-adenosyl methionine: DNA cytosine 5-methyl transferase from a Novikoff rat hepatoma cell line. Nucleic Acids Res 2:1685–1700

Tanaka M, Hibasami H, Nagei J, Ikeda T (1980) Effect of 5-azacytidine on DNA methylation in Ehrlich's ascites tumour cells. Aust J Exp Biol Med Sci 58:391–396

Turnbull JF, Adams RLP (1976) DNA methylase: purification from ascites cells and the effect of various DNA substrates on its activity. Nucleic Acids Res 3:677–695

Weintraub H, Larsen A, Groudine M (1981) α-Globin gene switching during the development of chicken embryos: expression and chromosome structure. Cell 24:333–344

Wigler M, Levy D, Perucho M (1981) The somatic replication of DNA methylation. Cell 24:33–40

Control of Maternal Inheritance
by DNA Methylation in Chlamydomonas

R. Sager, H. Sano, and C.T. Grabowy

1 Introduction

The maternal inheritance of chloroplast and mitochondrial genes is a fundamental property of the genetics of eukaryotic organisms. The first reports of chloroplast heredity were by Correns, who described strict maternal inheritance in *Mirabilis* and other plants (CORRENS 1909a, 1937) and by BAUR, who described maternal, paternal, and biparental patterns of transmission, all non-Mendelian, in the geranium plant, *Pelargonium* (BAUR 1909). In subsequent years, examples of maternally inherited chloroplast traits in higher plants have been widely reported and the phenomenon is now firmly established (SAGER 1972, 1977; GILLHAM 1978).

Throughout the history of this field, maternal inheritance has been the hallmark by which chloroplast and mitochondrial genes were distinguished from nuclear genes. The mechanism of maternal inheritance in higher organisms was widely assumed to result from the virtual absence of cytoplasm in the male gamete at the time of fertilization. Similarly, the occasional transmission of chloroplast traits from the male parent was assumed to result from transmission of occasional male proplastids or mitochondria in fertilization. Despite this "conventional wisdom", no direct evidence was available to support it, and some of the complex breeding results, e.g., with *Oenothera* and *Pelargonium*, cast doubt on the "physical exclusion" hypothesis as the sole mechanism of maternal inheritance (SAGER 1975; TILNEY-BASSETT 1975).

Dana-Farber Cancer Institute, Division of Cancer Genetics, 44 Binney Street, Boston, MA 02115, USA

Current Topics in Microbiology and Immunology, Vol. 108
© Springer-Verlag Berlin·Heidelberg 1984

Our discovery of maternal inheritance of chloroplast genes in *Chlamydomonas* (SAGER 1954) led us to a serious re-examination of the problem, since in this organism both parents contribute the total cell contents to the zygote. In the sexual cycle of *Chlamydomonas*, pairs of morphologically identical haploid cells of opposite mating type, mt^+ and mt^-, fuse to form the diploid zygote, which later undergoes meiosis and gives rise to four haploid zoospores, 2 mt^+ and 2 mt^-. Mating type is regulated by a nuclear gene or gene cluster. Chloroplast genes from the mt^+ parent are transmitted through the zygote to all four zoospores; the homologous chloroplast genes from the mt^- parent are not transmitted. This pattern of transmission is formally identical with the maternal inheritance of chloroplast genes in higher plants such as maize (*Rhoades* 1946); by analogy with the plant systems, we have considered the mt^+ parent the female and the mt^- parent the male. Exceptions to the rule of maternal inheritance occur with a frequency of about 0.1% in some strains of *Chlamydomonas*, and of up to 20% in other strains. In these exceptional zygotes, chloroplast genes are transmitted biparentally or rarely, from the paternal parent only (SAGER 1972; GILLHAM 1978).

The existence of maternal inheritance in *Chlamydomonas*, which occurs despite the equal contribution of cytoplasm from both parents, led us to seek an alternative to the physical exclusion hypothesis. Following the discovery of a high molecular weight species of DNA with a unique base composition in the chloroplast of *Chlamydomonas* (CHUN et al. 1963; SAGER and ISHIDA 1963) we turned our attention to an examination of chloroplast DNA transmission in the sexual cycle of *Chlamydomonas*. Our thinking was influenced by the process of modification and restriction of foreign DNA in bacteria, first described by LURIA (1953), and subsequently analyzed in several laboratories (see ARBER 1974; MESELSON et al. 1972; YUAN 1981). According to the modification-restriction hypothesis, foreign (e.g. phage) DNA entering a bacterial cell is degraded (i.e., restricted) by endonucleolytic attack at specific sequences on the DNA unless those sequences are protected by a secondary modification in DNA structure, later shown to be methylation. Methylation/restriction is an ideal mechanism for the destruction of one DNA genome in the presence of a homologous genome that was previously methylated. We therefore decided to search specifically for a methylation/restriction system in *Chlamydomonas*.

Over the past 10 years we have succeeded in obtaining several lines of evidence that support the methylation/restriction model as the molecular basis of maternal inheritance in *Chlamydomonas*. The model as initially formulated is shown in Fig. 1. It is interesting that a new and independent study of the disappearance of chloroplast DNA (chlDNA) after zygote formation, visualized by staining the DNA with DAPI, has confirmed our proposal that the chlDNA of one parent is degraded, while that of the other apparently remains intact, following zygote formation but before chloroplast fusion occurs (KUROIWA et al. 1982). This time course permits DNA degradation to occur in one chloroplast coincidental with or preceded by methylation in the other chloroplast, since the two chloroplasts are in sepa-

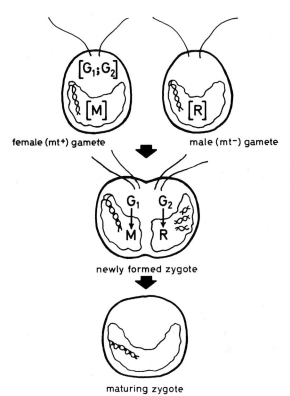

female (mt+) gamete   male (mt−) gamete

newly formed zygote

maturing zygote

Fig. 1. Postulated control of maternal inheritance of chloroplast DNA in *Chlamydomonas* by a methylation-restriction mechanism. Female (mt^+) gamete contains inactive modification enzyme *M* in chloroplast, and two regulatory substances, G_1 and G_2, in the cell sap. The male (mt^-) gamete contains inactive restriction enzyme *R* in its chloroplast. Before fusion of chloroplasts, the methylase is activated by G_1 to modify chloroplast DNA in the female chloroplast, and the restriction enzyme is activated by G_2 to degrade chloroplast DNA in the male chloroplast. The two chloroplasts then fuse, and only the chloroplast DNA from the female parent is available for replication. (Nuclei are not shown for sake of clarity.) (From SAGER and RAMANIS 1973)

rate compartments in the period preceding chloroplast fusion. In the years since our model was first proposed, we have established that DNA methylation occurs not only after zygote formation, but also preceding zygote formation during gametogenesis in female gametes but not in males. Thus, the chlDNA of female origin may enter the zygote already protected against restriction.

This chapter summarizes the evidence that regulated methylation of chlDNA occurs at specific stages in the life cycle of *Chlamydomonas*, and plays a key role in the control of maternal transmission of chloroplast genes.

2 Background Data

The first evidence that methylation and restriction regulate the transmission of chlDNA came from studies showing that chlDNA of female origin, present in the zygote and identified by an ^{15}N prelabel, has undergone a density shift in comparison to chlDNA from vegetative cells and consistent with methylation. In contrast the homologous DNA of male origin does not show the shift. In the same studies it was also shown that prelabeled chlDNA of male origin disappeared soon after zygote formation, whereas the chlDNA of female origin was present and identifiable by its label (SAGER and LANE 1972). These findings were subsequently confirmed and amplified using ^{3}H-thymidine to follow the fate of chlDNAs in zygotes after fusions in reciprocal crosses of labeled with unlabeled parents. Further, the identification of 5-methylcytosine (5mC) in chlDNA was achieved by prelabeling with (G-^{3}H)deoxycytidine (BURTON et al. 1979). As shown in Figure 2, 5mC was detected in HPLC separations of bases from chlDNA of zygotes only when the female was prelabeled. Methylation was essentially complete within 6 h after fusion. It was also shown that no other bases were methylated, and no 6-methyladenine (6mA) was detected after prelabeling with ^{3}H-adenine. Thus, both the preferential cytosine methylation of chlDNA of female origin and degradation of the homologous chlDNA of male origin were established with the methods available at the time.

A DNA endonuclease with some of the properties of a restriction enzyme was isolated and partially characterized (BURTON et al. 1977). In collaboration with R. Roberts, using the same methodology which was applied so successfully to the detection and purification of bacterial restriction endonucleases, we found a unique deoxyribonuclease in extracts of zygotes and vegetative cells of *Chlamydomonas*. Enzyme activity in crude extracts of *Chlamydomonas* cleaved adeno-2 DNA, producing discrete fragments upon electrophoresis in agarose. However, the fragments were not equimolar, and were shown to result from single strand nicks and overlapping gaps made by the enzyme at specific sites. Studies with synthetic polynucleotides showed that the enzyme cleaved only at sites containing T, but the 5' termini present in cleaved and gapped adeno-2 DNA included about 25% G as well as 66% T residues. To account for these results we proposed that the enzyme initially attacks at a site containing T but that a second cut, leading to excision of a single stranded fragment, might occur at a site with different specificity (BURTON et al. 1977). Thus this enzyme, the first site-specific endonuclease described from a eukaryote, resembles a type I endonuclease of the kind described originally by LINN (ARBER and LINN 1969) and by YUAN and MESELSON (MESELSON et al. 1972).

Efforts to purify this enzyme have not succeeded thus far. It should be noted that enzymes with similar properties have been found in mammalian tissues (J.J. Maio, personal communication) and in Epstein-Barr virus-producing human lymphoblastoid cells (CLOUGH 1980). The existence of these enzymes in mammalian cells suggests that site-specific DNA cleavage and degradation may play a role as yet undefined in normal development.

3 Methylation of Gametes

When restriction enzymes became available as analytical tools, we compared restriction fragment patterns in vegetative and gametic chlDNAs from males and females to look for differential methylation. Initially *MspI/HpaII* patterns were compared (ROYER and SAGER 1979). As suspected, chlDNA from

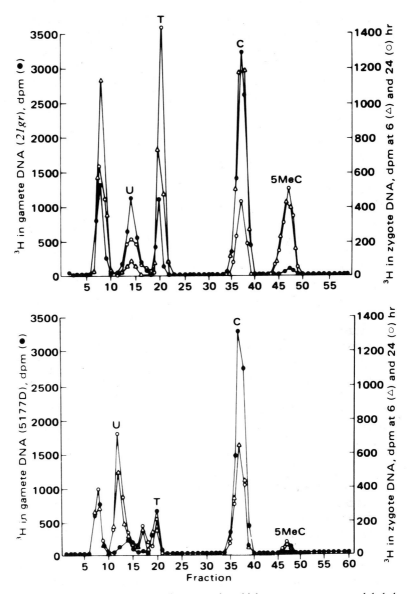

Fig. 2. Chloroplast DNAs from gametes and zygotes in which one parent was prelabeled with deoxycytidine. One strain was prelabeled with [G-³H]deoxycytidine (50 µCi/ml), gametes were prepared and fused with unlabeled gametes of the other strain, and zygotes were sampled at 6 and 24 h. Chloroplast DNAs were purified and hydrolyzed with formic acid, and bases were separated on HPLC Aminex A-6 column. Positions of marker bases are shown: *U*, uracil; *T*, thymine; *C*, cytosine; *5MeC*, 5-methylcytosine. Peak at left is void volume, containing [³H]deoxyribose. (Upper) Strain *21gr* labeled. (Lower) Strain *5177D* labeled. Identification of uracil is uncertain

female gametes was very resistant to *HpaII* digestion, whereas homologous chlDNAs from male gametes, and from both male and female vegetative cells, were uniformly sensitive to *HpaII*, giving fragment patterns very similar to those seen with *MspI*. These results showed dramatically that methylation of the internal cytosine in CCGG sites had occurred in female gametes, but not in males, during the 24 h period of gametogenesis.

Since examination of CCGG sites reveals only a small fraction of the total methylation, other methods are required to provide a more complete analysis of the extent and location of methylation occurring during gametogenesis, as well as of the further methylation occurring in zygotes, which we previously found. Precise sequence localization of 5mC can be established by the method of MAXAM and GILBERT (1977), but only with uncloned DNA. Quantitative methods such as paper or liquid chromatography or mass spectroscopy provide no information on location.

A new and powerful method was developed in our laboratory for quantification and localization of 5mC, in restriction fragments, namely the use of antibodies directed against 5mC in DNA (SANO et al. 1980). With this method, potentially every methylated cytosine can be detected and localized by restriction site mapping. The sensitivity of the method is such that single methylated bases can be recognized and as little as 0.02 pmol 5mC can be detected. The level of detection, of course, varies with the radioactivity of the probe.

The method is straightforward. DNA restriction fragments are transferred from agarose gels to nitrocellulose paper where they are immobilized, and incubated first with the rabbit antibody against 5mC, and then with goat anti-rabbit IgG which has been prelabeled with ^{125}I. Restriction fragments containing 5mC are visualized by autoradiography.

The first demonstration of methylation in *EcoRI* digested chlDNAs from female gametes and zygotes by this method is shown in Fig. 3. The fragments are identified by ethidium bromide (EB) fluorescence (Fig. 3A), and partial resistance to *EcoRI* digestion can be seen in lanes *c* and *e*, female gamete and zygote DNAs respectively. In Fig. 3B, the extensive methylation of these two DNAs is evident in contrast to the absence of detectable methylation in the other lanes.

In a subsequent more detailed study, not only *EcoRI* but also *MspI* and *HpaII* fragment patterns were examined, both by EB fluorescence and by antibody binding (SAGER et al. 1981). We also examined the methylation patterns of the *mat-1* mutant. The *mat-1* mutation was discovered in a male (*mt$^-$*) population (SAGER and RAMANIS 1974). It does not segregate from *mt$^-$* in crosses, indicating close linkage to the mating type locus. When *mat-1 mt$^-$* mutant cells are mated with wild-type *mt$^+$*, chloroplast genetic markers are transmitted from both parents. Thus, the *mat-1* mutation, carried by males, converts maternal to biparental inheritance.

As shown in Fig. 4, we compared the restriction fragment patterns from chlDNA digested with *MapI* and *HpaII* from two wild-type females (*mt$^+$*) lines, *21gr* and *12-636-6*, and from two males, *mat-1* and the wildtype male *5177D*. This figure illustrates a number of points. The *MspI* and *HpaII*

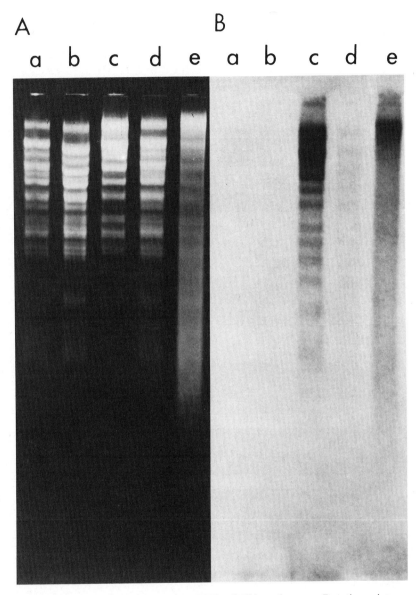

Fig. 3A, B. Identification of m^{5}C in chloroplast DNA of *Chlamydomonas*. Reaction mixtures (50 µl) containing 0.1 M Tris-HCl (pH 7.6), 50 mM NaCl, 5 mM MgCl$_2$, 2 µg of DNA, and 25 units of EcoRI were incubated at 37° C for 6 h. Samples were electrophoresed at 100 V for 17 h. DNAs were visualized by ethidium bromide staining (**A**) and m^{5}C was identified by antibody binding (**B**). Lanes: *a*, DNA from female vegetative cells; *b*, DNA from male vegetative cells; *c*, DNA from female gamete cells; *d*, DNA from male gamete cells; and *e*, DNA from zygote cells

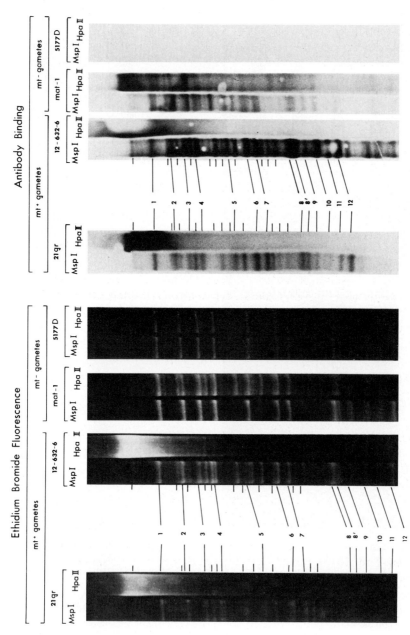

Fig. 4. *MspI* and *HpaII* digests of chloroplast DNA from gametes. Four cell lines are compared: *21gr* is wild-type *mt*$^+$ (female); *12-632-6* is spiromycin-resistant *mt*$^+$ (female); *5177D* is streptomycin-resistant *mt*$^-$ (male); *mat-1* is a mutant showing biparental instead of maternal inheritance, *mt*$^-$ (male). The restriction fragment patterns revealed by ethidium bromide fluorescence of *MspI*- and *HpaII*-restricted chloroplast DNA are shown in the first four pairs of lanes; and the corresponding patterns revealed by ^{125}I-labeled goat antirabbit IgG bound to rabbit antibody against 5-methyl cytosine are shown in the second four pairs of lanes. Numbering follows ROCHAIX (1978); unnumbered bands are not seen in chloroplast DNA from vegetative cells

patterns compared by EB fluorescence show that extensive methylation of the internal C of CCGG has occurred during gametogenesis in the female but not in the wild-type male, *5177D*. However, the *HpaII* pattern in *mat-1 mt⁻* gametes is intermediate, showing extra bands in the *HpaII* lane but not extensive methylation. In the autoradiographs that show the extent of methylation in each band it is evident that there is differential methylation of different *MspI* fragments in the *21gr* lane (wild-type female). Thus methylation is not random in the sense that it is not proportional to the size of the fragment. Furthermore, extra fragments are present in the *MspI* digest of wild-type female gamete DNA, but not present in the corresponding vegetative DNA. These bands are the result of methylation of the 5′ cytosine of CCGG. No methylation whatsoever was detected in the wild-type male gametes (*5177D*).

The results with the *mat-1* mutant are of particular interest. Differences in the *MspI* and HpaII patterns in the autoradiographs confirm what is seen in the gels by EB fluorescence. The *HpaII* digest of *mat-1* DNA resembles the *MspI* digests of female wild-type DNA. Thus, methylation may be proceeding more slowly in the *mat-1* mutant than in the wild-type female gametes. Support for this idea was presented in a study of the restriction fragment patterns of chlDNA, sampled at intervals during gametogenesis of female wild-type cells (SAGER et al. 1981). The extent of methylation of the wild type at 4 and 8 h of gametogenesis resembled that of the *mat-1* mutant at 24 h. Thus, chlDNA of *mat-1* males is methylated at a slower rate than in the wild type females.

This results provided the first evidence that the *rate* of methylation is under genetic control. Further evidence has come from studies of the *me-1* mutant to be discussed now.

In a recent report, BOLEN et al. (1982) have described a mutant of *Chlamydomonas* called *me-1* in which the chlDNA of vegetative cells, both male and female, is extensively methylated. The total methylation determined by HPLC was reported as 7 mol% of total bases and 37% of the cytosine residues. This extensive methylation has had no reported effect on the phenotype of the cells or on maternal inheritance. With respect to the phenotype, this result suggests that methylation of chlDNA does not influence gene expression, perhaps because the effects of methylation on eukaryotic gene transcription require chromosomal proteins and nucleosome structure, absent in chlDNA. This question will be addressed elsewhere. However, the fact that extensive methylation of chlDNA in male as well as female cells of the *me-1* mutant did not alter maternal inheritance of chloroplast genes, will be addressed here.

We have examined the extent and pattern of chlDNA methylation in the *me-1* mutant cells of both mating types, kindly provided by Dr. N.W. Gillham. We have compared restriction fragment patterns after agarose gel electrophoresis of chlDNAs from mutant vegetative cells and gametes with those from the wild type, using a set of 32 restriction enzymes of which 17 were methylation-sensitive in this system (SAGER and GRABOWY 1983). We have found that additional methylation occurs during gametogenesis in the *me-1* female (mt^+) but not in the male (mt^-). Thus, gamete specific, mating-type specific, methylation occurs in the *me-1* mutant as in the wild-type, consistent with our methylation-restriction model.

As an example of the results, restriction fragment patterns are shown in Fig. 5, representing chlDNAs from wild-type and mutant vegetative cells and gametes, each digested with *Msp*I (Fig. 5a), *Hinf*I (Fig. 5b) or *Hae*III (Fig. 5c). In each comparison, additional methylation was seen in the *me-1* mutant female gametes (mt^+) compared to the vegetative cells but not in the male (mt^-). Thus, the mutant female vegetative, male vegetative, and male gamete were alike, whereas the mutant female gametes were much more highly methylated than the vegetative cells from which they were derived. It is evident from these comparisons that increased methylation of chlDNA of the female gametes corresponds with the genetic evidence of maternal inheritance in the *me-1* mutant just as in the wild type.

In addition, we have examined total 5mC content of cells from various stages (GRABOWY and SAGER, ms. in prep.). We have found that populations of *me-1* mt^+ vegetative cells reaching stationery phase of growth ($3-4 \times 10^6$ cells/ml) have higher methylation levels (ca. 7 mol% of 5mC in total bases) than exponential populations which contain about 5 mol% 5mC. In wild-type mt^+ cells, methylation increased during gametogenesis. After 24 h, the 5mC content was about 4 mol%, whereas after 48 h, this value increased to 6 mol%, in the range of *me-1* mt^+ vegetative cells. On the basis of these data, as well as comparative restriction fragment patterns, we estimate that the extent and specificity of methylation in the *me-1* mutant vegetative cells is similar to that of the wild type female gametes at specific times in their cycles.

Our results are not in disagreement with the data presented by *Bolen* et al. (1982), except that our mt^+ gametes show much more extensive *Hpa*II resistant methylation than is reported by them. Unfortunately, they did not compare restriction fragment patterns of vegetative cells and gametes of the two mating types, and thus they failed to observe the mating-type specific increase in methylation that occurs in gametogenesis in the *me-1* female cells.

In summary, all the evidence so far available supports the role of cytosine methylation of chlDNA in the molecular control of maternal inheritance. The specific sites that are protected by methylation and cleaved by restriction have not yet been identified. It is hoped that the differential methylation patterns in wild-type and in *mat-1* and *me-1* mutants may provide material from which the identity of these sites may be deduced.

4 Kinetic Studies of Methylation in Gametogenesis and Its Reversal

The time course of methylation during gametogenesis was examined with chlDNA from wild-type female cells, restricted with the enzymes *Msp*I, *Hpa*II, and *Eco*RI. Cell samples were taken 4, 8 and 16 h after cells had been suspended in a medium free of available nitrogen source to initiate

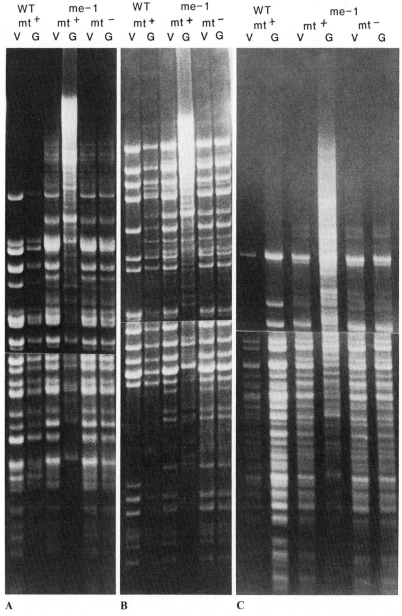

Fig. 5 A–C. Restriction fragment patterns of chloroplast DNA digested with (**A**) HaeIII, (**B**) MspI, (**C**) HinfI. *V*, vegetative; *G*, gamete; *mt⁺*, female; *mt⁻*, male; *WT*, wild type; *me-1*, methylated mutant

gametogenesis. We found that methylation increased progressively with time. Some *HpaII* restriction resistance was already evident at 4 h. By 16 h the *MspI* pattern clearly showed extra bands, indicating the occurrence of methylation at the 5' cytosine of CCGG. The methylation of the internal

C detected in *HpaII* digests was extensive, with much of it occurring between 8 and 16 h. A few extra bands were also seen in the *EcoRI* lane at 16 h, indicating that *EcoRI* is a methylation-sensitive enzyme, and consistent with our previous report of resistance of zygotic chlDNA to *EcoRI* digestion (ROYER and SAGER 1979).

We had previously shown (SAGER and GRANICK 1954) that gametes can be dedifferentiated by transferring them into a standard medium containing an available nitrogen source; they lose mating ability and reenter the cell cycle. In recent studies we have found that more than 90% of mating efficiency is lost in 7 h, before cells have divided and before methylation patterns have changed (SANO et al. manuscript in preparation). Thus, the mating reaction *per se* is evidently not directly determined by the level of methylation.

The extent of methylation has been determined during gametic differentiation and dedifferentiation by two methods: (a) reversed phase HPLC to separate deoxyribonucleosides (KUO et al. 1980), and (b) nick translation (four reactions each with a different nucleotide substrate) followed by enzymatic hydrolysis generating 3′-dNMPs; the amount of 5mC was estimated from ^{32}P-radioactivity after two-dimensional thin layer chromatography (SANO et al. ms. in prep.).

The results agree in showing that the final extent of methylation in 24 h gametes is about 22% of the total cytosines, or 4 mol% of DNA. When the extent of methylation was compared with the amount of cell division over a period of 4 days (six population doublings) after adding a nitrogen source to gametes, it was found that the methylation decreased more slowly than had been expected on the basis of dilution. Since the level of 200-K DNA methyltransferase activity (see Sect. 5) decreased proportionately with cell division, the presence of some active enzyme during dedifferentiation could account for the slow kinetics of decreased methylation. Thus, these results do not support a process of active demethylation, but rather suggest a passive process of demethylation by dilution of the methyltransferase during cell division.

5 DNA Methyltransferase

DNA methylation is an enzymatic reaction catalyzed by DNA methyltransferase(s) (EC 2.1.1.37). In a preliminary search for methyltransferase activity in *Chlamydomonas*, we found that vegetative cells contained an enzyme which catalyzed the transfer of methyl groups from S-adenosylmethionine to the 5 position of cytosine residues in double stranded DNA. The enzyme was purified up to 310-fold by high-speed centrifugation, DEAE- and phospho-cellulose chromatography. The native enzyme with mol.wt. of 58000 accepted dsDNA preferentially as the substrate, and introduced methyl groups into sequences containing 5′d(T-5mC-pU)3′ (SANO and SAGER 1980). However, since DNA from chloroplasts and nuclei of vegetative cells is not methylated, the presence of this highly active enzyme in extracts of vegetative cells was puzzling.

Table 1. DNA methyltransfere activity at various stages in life cycle of Chlamydomonas

Stage in life cycle	No. of expts.	Enzyme, units per 10^9 cells (% of total)		
		Total	M_r 60000	M_r 200000
mt^+ vegetative	6	3.4	2.8 (86)	0.6 (14)
mt^+ gamete	2	3.1	1.7 (55)	1.4 (45)
mt^- gamete	2	1.8	1.8 (100)	–
mt^- mat-1 gamete	2	6.7	5.1 (76)	1.6 (24)
Zygote	3	22.8	–	22.8 (100)

Total activity was estimated from elution profiles of enzymes from DEAE-cellulose column chromatography in replicate experiments. One unit of enzyme activity is defined as 1 pmol of methyl group incorporated into DNA per 60 min at 37° C in the presence of *Micrococcus luteus* DNA at 70 µg/ml

This question was partially resolved in a further study of DNA methyl-transferase activity from gametes and zygotes (SANO et al. 1981). Using methods similar to those mentioned above, we have isolated two forms of a methyltransferase, a 60-K form and 200-K form. The 60-K form of the enzyme is present in vegetative cells and gametes of both mating types, and probably corresponds to the 58-K enzyme previously described (SANO and SAGER 1980). In enzyme preparations from mt^+ gametes, however, a 200-K form is also present. Similarly, both the 60-K and 200-K forms are present in *mat-1 mt^-* gametes. Zygotes contain only the 200-K form. The quantitative estimation of each enzyme at various stages in the life cycle is summarized in Table 1. It is clear that the 200-K form is present only in the cells in which methylation occurs.

We postulate that the 200-K protein is a multimer of the 60-K form and that only the 200-K form is active in vivo, for the following reasons (SANO et al. 1981). (a) The two forms show the same methylation pattern in vitro, introducing methyl groups preferentially into d(C,G); (b) on glycerol density gradients the 200-K component from mt^+ gametes gives both 60-K and 200-K peaks.

A more detailed analysis of methylation by the 200-K enzyme *in vitro* revealed that methylation occurred randomly at CA, CG, CT, and CC. When *Micrococcus luteus* DNA was used as the substrate, the methylation pattern paralleled the relative frequencies of these doublets. However, the synthetic polymer poly(dC)·poly(dG), was not methylated by this enzyme, suggesting that specific sequences and/or a special DNA conformation may be necessary for the enzyme to bind to the DNA substrate. In this sense, the 200-K enzyme may be similar to rat liver DNA methyltransferase (DRA-HOVSKY and MORRIS 1971) and topbacterial type I restriction enzymes (YUAN 1981).

The mode of methylation by the 200-K enzyme was examined with nonmethylated (vegetative), hemimethylated (reannealed DNA from vegetative and gamete cells), and methylated (gametes) DNA substrates (SANO et al. 1981). The methylation velocity at a fixed amount of substrate DNA was highest with hemimethylated DNA, followed by nonmethylated DNA and slowest with methylated DNA. Thus, the 200-K enzyme methylated hemi-methylated sites in DNA more rapidly than it did nonmethylated sites. These data strongly suggest that the 200-K enzyme acts as both initiation and maintenance methyltransferase.

6 Concluding Remarks

The methylation-restriction hypothesis accounts for the mechanism of maternal inheritance of chlDNA and chloroplast genes in *Chlamydomonas* with increasing precision as studies of this system progress. Using wild-type cells, we have described the overall process of methylation that occurs during gametogenesis in female (mt^+) cells but not in males (mt^-). We have demonstrated this differential methylation by several methods: HPLC identification of 5mC of female origin in zygotes (BURTON et al. 1979; restriction-fragment pattern differences using methylation-sensitive enzymes (ROYER and SAGER 1979; SAGER et al. 1981; SAGER and GRABOWY 1983); presence of 5mC detected in restriction fragments by anti-5mC antibodies (SANO et al. 1980); and presence of a 200-K methyltransferase only in stages of the life cycle during which chlDNA methylation occurs (SANO et al. 1981).

Further light on the process has been cast by the examination of two mutants that alter wild-type methylation patterns: *mat-1* and *me-1*. The *mat-1* mutation, originally detected in male cells and closely linked to the mating type locus, is responsible for a shift from maternal to biparental inheritance. At the molecular level, *mat-1* mutant gametes exhibit partial methylation of chlDNA and contain the 200-K form of the methylase. A comparison of methylation in restriction fragments of wild-type and *mat-1* gametes by the antibody binding method indicates that 24 h *mat-1* gametes resemble early stages in the time course of wild-type gametogenesis (SAGER et al. 1981). This result, as well as the decreased content of the 200-K enzyme in *mat-1* compared with the wild type, suggests that the rate of appearance of the 200-K enzyme is slower in the *mat-1* mutant than in wild-type female gametes; and further suggests that the *mat-1* gene regulates assembly of the 200-K enzyme.

Our studies with the *me-1* mutant cell lines (SAGER and GRABOWY 1983) briefly described here and depicted in Fig. 5, show that *me-1* cells exhibit increased methylation of the female cells but not of the male cells during gametogenesis, just as in the wild type. However, the baseline level of methylation, upon which gametic methylation is introduced, is much higher in *me-1* cells than in the wild type. These results confirm and strengthen the generality of our previous findings: differential methylation of chlDNA in males and females is consistent with maternal or, in *mat-1* mutants, biparental inheritance of chloroplast genes.

The question remains unanswered whether the methylation in *me-1* mutant vegetative cells is carried out by the same enzyme that is activated in gametogenesis of wild-type females, and of *mat-1* mutant males, or whether more than one enzyme is involved. We recently proposed a two-enzyme hypothesis, according to which different enzymes were considered responsible for the methylation occurring in vegetative *me-1* cells, and in gamtes (SAGER and GRABOWY 1983). However, a simpler hypothesis can be formulated, consistent with the data so far available, that involves the regulation of a single enzyme.

The biological data require that the methylation occurring in *me-1* vegetative cells does not confer resistance to degradation in zygotes, whereas the subsequent methylation occurring in gametes *does* confer resistance. The one-enzyme hypothesis is based on the evidence reviewed above showing that methyl transferase activity was detected in two molecular weight fractions from DEAE-cellulose columns, a 60-K and a 200-K form. Biological methylating activity correlates with the presence of the 200-K form, whereas in non-methylated cells the 60-K form is present and active in in vitro assays, but presumably inactive in vivo.

According to the one-enzyme hypothesis, it is postulated that the 60-K form is inactive in wild-type vegetative cells, but is activated in the *me-1* mutant. Thus, it is proposed that the *me-1* mutation is regulatory, either removing an inhibitor or modifying the enzyme. A different mode of regulation is proposed for gametogenesis, involving assembly of the 60-K subunits into a 200-K multimer, regulated by the *mat-1* gene at the mating type locus, as previously discussed (SANO et al. 1981). The new element in this latter proposal is the requirement that the site specificity of methylation by the 200-K enzyme is different from that of the activated 60-K enzyme, since methylation by the 200-K enzyme confers resistance to degradation occurring in the zygote, whereas vegetative cell methylation in *me-1* mutants does not. Many precedents exist for specificity changes resulting from allosteric interactions in enzyme assembly that could account for the observed changes in restriction sensitivity. Thus, the one-enzyme hypothesis is based on reasonable speculation, but awaits experimental test.

References

Arber W (1974) DNA modification and restriction. Prog Nucl Acid Mol Biol 14:1–37
Arber W, Linn S (1969) DNA modification and restriction. Rev Biochem 38:467
Baur E (1909) Das Wesen und die Ehrlichkeitsverhaltnisse der "Varietates albomarginatae hort" von Pelargonium zonale. Z Vererbungslehre 1:330–351
Bolen PL, Grant DM, Swinton D, Boynton JE, Gillham NW (1982) Extensive methylation of chloroplast DNA by a nuclear gene mutation does not affect chloroplast gene transmission in Chlamydomonas. Cell 28:335–343
Burton WG, Roberts RJ, Myers PA, Sager R (1977) A site-specific single strand endonuclease from the eukaryote *Chlamydomonas*. Proc Natl Acad Sci USA 74:2687–2691
Burton WG, Grabowy C, Sager R (1979) The role of methylation in the modification and restriction of chloroplast DNA in *Chlamydomonas*. Proc Natl Acad Sci USA 76:1390–1394
Chun EHL, Vaughan MH, Rich A (1963) The isolation and characterization of DNA associated with chloroplast preparations. J Mol Biol 7:130–141
Clough W (1980) An endonuclease isolated from Epstein-Barr virus-producing human lymphoblastoid cells. Proc Natl Acad Sci USA 77:6194–6198
Correns C (1909a) Vererbungsversuche mit blass (gelb) grunen und buntblattrigen Sippen bei *Mirabilis*, *Urtica* and *Lunaria*. Z Vererbungslehre 1:291–329

Correns C (1909b) Zur Kenntnis der Rolle von Kern und Plasma bei der Vererbung. Z Verer-
    bungslehre 2:331–340
Correns C (1937) Nicht Mendelnde Vererbung. Borntraeger, Berlin
Drahovsky D, Morris NR (1971) Mechanism of action of rat liver DNA methylase. J Mol
    Biol 57:475–489
Gillham NW (1978) Organelle heredity. Raven, New York
Kuo KC, McCune RA, Gehrke (1980) Quantitative reversed phase high performance liquid
    chromatographic determination of major and modified deoxyribonucleosides in DNA. Nucl
    Acids Res 8:4763–4776
Kuroiwa T, Kawano S, Nishibayashi S (1982) Epifluorescent microscopic evidence for maternal
    inheritance of chloroplast DNA. Nature 298:481–483
Luria SE (1953) Host induced modifications of viruses. Cold Spring Harbor Symp Quant
    Biol 18:237
Maxam AM, Gilbert W (1977) A new method for DNA sequencing. Proc Natl Acad Sci
    USA 74:560–564
Meselson M, Yuan R, Heywood J (1972) Restriction and modification of DNA. Ann Rev
    Biochem 41:447–466
Rhoades MM (1946) Plasmid mutations. Cold Spring Harbor Symp Quant Biol 11:202–207
Royer HD, Sager R (1979) Methylation of chloroplast DNA in the life cycle of *Chlamydomonas*.
    Proc Natl Acad Sci USA 76:5794–5798
Sager R (1954) Mendelian and non-Mendelian inheritance of streptomycin resistance in *Chla-
    mydomonas reinhardi*. Proc Natl Acad Sci USA 40:356–363
Sager R (1972) Cytoplasmic genes and organelles. Academic, New York
Sager R (1975) Patterns of inheritance of organelle genomes: molecular basis and evolutionary
    significance. In: Genetics and biogenesis of mitochondria and chloroplasts. Birky CW,
    Perlman PS, Byers TJ, eds, Ohio State University Press, Ohio
Sager R (1977) Genetic analysis of chloroplast DNA in *Chlamydomonas*. Adv Genetics
    19:287–340
Sager R, Granick S (1954) Nutritional control of sexuality in *Chlamydomonas reinhardi*. J
    Gen Physiol 37:729–742
Sager R, Ishida MR (1963) Chloroplast DNA in *Chlamydomonas*. Proc Natl Acad Sci USA
    50:725–730
Sager R, Lane D (1972) Molecular basis of maternal inheritance. Proc Natl Acad Sci USA
    69:2410–2413
Sager R, Ramanis Z (1973) The mechanism of maternal inheritance in *Chlamydomonas*. Bio-
    chemical and genetic studies. Theor Appl Genet 43:101–108
Sager R, Ramanis Z (1974) Mutations that alter the transmission of chloroplast genes in
    Chlamydomonas. Proc Natl Acad Sci USA 71:4698–4702
Sano H, Sager R (1980) Deoxyribonucleic acid methyltransferase from the eukaryote *Chlamy-
    domonas reinhardi*. Eur J Biochem 150:471–480
Sager R, Grabowy C (1983) Regulation of maternal inheritance by differential methylation
    of chloroplast DNA in *me-1* mutant of *Chlamydomonas*. Proc Natl Acad Sci USA
    80:3025–3029
Sager R, Grabowy C, Sano H (1981) The *mat-1* gene in *Chlamydomonas* regulates DNA
    methylation during gametogenesis. Cell 24:41–47
Sano H, Royer HD, Sager R (1980) Identification of 5-methylcytosine in DNA fragments
    immobilized on nitrocellulose paper. Proc Natl Acad Sci USA 77:3581–3585
Sano H, Grabowy C, Sager R (1981) Differential activity of DNA methyltransferase in the
    life cycle of *Chlamydomonas reinhardi*. Proc Natl Acad Sci USA 78:3118–3122
Sano H, Grabowy C, Sager R (1982) Loss of chlDNA methylation during dedifferentiation
    of gametes in *Chlamydomonas* (Manuscript in preparation)
Tilney-Bassett RAE (1975) Genetics of variegated plants. In: Genetics and biogenesis of mito-
    chondria and chloroplasts. Birky CW, Perlman PS, Byers TJ, eds. Ohio State University
    Press, Ohio
Yuan R (1981) Structure and mechanism of multifunctional restriction endonucleases. Ann
    Rev Biochem 50:285–315

Note Added in Proof to the Chapter by Günthert/Trautner, p. 11

Since the submission of this manuscript, we have established the nucleotide sequence of the SPR MTase gene (BUHK et al., submitted for publication). The gene has 1317 base pairs. The mol. wt. of the MTase predicted from the sequence is 49900. This is in good agreement with the mol. wt. of the chromatographically purified MTase and with the gene product identified by minicell technique. Base changes responsible for mutations affecting SPR methylation (see p. 17 and 18) have been localized at various positions within the 1317 base pairs of the gene.

Subject Index

Current Topics in Microbiology and Immunology

Editors:
M. Cooper, W. Henle,
P. H. Hofschneider,
H. Koprowski, F. Melchers,
R. Rott, H. G. Schweiger,
P. K. Vogt, R. Zinkernagel

Springer-Verlag
Berlin
Heidelberg
New York
Tokyo

Volume 98

Retrovirus Genes in Lymphocyte Function and Growth

Editors: **E. Wecker, I. Horak**

1982. 8 figures. VIII, 142 pages
ISBN 3-540-11225-1

"This timely little volume ... (is a) coherent and crisp publication ... summarizing much of the important work and thinking in the field (of RNA tumour virus) ... welcomed by many who want a competent overview or an update." *Immunology Today*

Volume 100

T Cell Hybridomas

A Workshop at the Basel Institute for Immunolgy
Organized and edited by H. V. Boehmer, W. Haas,
G. Köhler, F. Melchers, J. Zeuthen
With the collaboration of S. Buser-Boyd

1982. 52 figures. XI, 262 pages
ISBN 3-540-11535-8

To help evaluate the potential of T cell hybridomas for further understanding of the functioning of the immune system as well as for practical purposes, leading experts were invited to a workshop with the aim of publishing their contributions so that their approach and their techniques used to obtain, maintain, and analyze T cell hybridomas, the difficulties they encountered and how they were overcome (or not) be accessible to all interested. The reports present the state-of-the-art on T cell hybridomas research.

Volume 104

New Developments in Diagnostic Virology

Editor: **P. A. Bachmann**

1983. 117 figures. XII, 330 pages
ISBN 3-540-12171-4

The Molecular Biology of Adenoviruses 1

30 Years of Adenovirus Research 1953–1983

Edited by **Walter Doerfler**

1984. 69 figures. Approx. 240 pages
(Current Topics in Microbiology and
Immunology, Volume 109)
ISBN 3-540-13034-9

The Molecular Biology of Adenoviruses 2

30 Years of Adenovirus Research 1953–1983

Edited by **Walter Doerfler**

1984.
(Current Topics in Microbiology and
Immunology, Volume 110)
ISBN 3-540-13127-2
In preparation

The Molecular Biology of Adenoviruses 3

Edited by **Walter Doerfler**

1984.
(Current Topics in Microbiology and
Immunology, Volume 111)
ISBN 3-540-13138-8
In preparation

Springer-Verlag
Berlin
Heidelberg
New York
Tokyo